THE KEY TO PSYCHOTHERAPY

Understanding the Self-Created Individual

Second Edition, Revised and Expanded

Robert L. Powers and Jane Griffith

Adlerian Psychology Associates, Ltd., Publisher
Port Townsend, Washington 98368-0036

Printed in the United States of America

Library of Congress Control Number: 2010910816

E-book © 2012; ISBN: 978-0-918287-15-1
Paperback © 2012; ISBN: 978-0-918287-18-2
Hardcover © 2012; ISBN: 978-0-918287-19-9

Second edition, revised and expanded
Original work published 1987 titled,
  Understanding Life-Style: The Psycho-Clarity Process

Queries: info@adlerpsy.com
Orders: www.adlerianpsychologyassociates.com

THE KEY TO PSYCHOTHERAPY

Understanding the
Self-Created Individual

BY THE SAME AUTHORS

The Lexicon of Adlerian Psychology: 106 Terms Associated with the Individual Psychology of Alfred Adler (2nd Rev. ed., 2007; E-book, 2010)

The Individual Psychology Client Workbook with Supplements (3rd Rev. ed., 2012; E-book, 2012)

IN MEMORIAM

Thomas Philip Powers

1934 - 2008

A loyal son and loving husband, father, brother,
and friend. As a devoted labor lawyer, Tom was
brilliant, scrappy, brave, and self-sacrificing on
behalf of workers abused by disregard or deceit.

He was a model of the fellow man.

Production of this book was made possible in part by a gift in Tom's honor.

PREFACE TO THE FIRST EDITION

Our plan was to produce a small manual for users of *The Individual Psychology Client Workbook (IPCW)* (1986). We soon recognized that we were writing a book. The further we went the more the book seemed to require and the more it grew, until we finished (or, stopped) with the text you now have in hand. It is much more detailed and extensive than we pictured it to be in our original vision; however, it remains a text to accompany the *IPCW*.

Our book did not grow into a text on how to do psychotherapy; we assume that our readers are already competent for this work, or are otherwise preparing for it. We also assume a general understanding of Individual Psychology and its clinical applications. Now that we can see what we have written however, we believe that we have produced a useful and instructive introduction to Adlerian theory and practice for people trained in other schools of thought. If this belief turns out to be justified, and if our work inspires further study of Individual Psychology, all the better.

Adler wrote and taught during more than a third of the 20th century; his ideas went on developing as he worked. Our references to his writings almost always cite one of the trilogy of works collected and edited by Heinz L. and Rowena R. Ansbacher. These books have made the full range of Adler's thought accessible to students; researchers will also find in them any information they may require concerning the original writings.

We offer the psychoclarity process [Chapter Two] as a method not only for *Understanding Lifestyle,* but also for encouraging people. It is a way for clinicians to rediscover with their clients a sense Adlerians call social interest [community feeling], the sense of our being responsible for shaping the world we share.

Robert L. Powers
Jane Griffith

May, 1987
Chicago, Illinois

PREFACE TO THE SECOND EDITION

Welcome to the second edition, revised and expanded, of a well-used text that follows in the school of Alfred Adler. His work instructed and inspired us, and his methods are reflected here. With him, we invoke the artistic quality of personality development, focusing on how the self-created person shapes outlook, attitude, and patterns of movement among others. Summing up his "fundamental views" in 1935, Adler said,

> Life (and all psychological expressions as part of life) moves ever toward overcoming, toward perfection, toward superiority, toward success. You cannot train or condition a living being for defeat. But what an individual thinks or feels as success (as an acceptable goal) is unique with him. In our experience we have found that each individual has a different meaning of, and attitude toward what constitutes success. Therefore a human being cannot be typified or classified. We believe that the parsimony of language causes many scientists to come to mistaken conclusions, to believe in types, entities, and racial qualities. Individual Psychology recognizes . . . that each individual must be studied in the light of his own peculiar development. To present the individual understandably, in words, requires an extensive reviewing of all his facets. Yet too often psychologists are tempted away from this recognition to take the easier but unfruitful roads of classification. That is a temptation to which, in practical work, we must never yield (1956, p. 167).

Here again is the step-by-step guide for presenting "the individual understandably, in words" to your client, with all the liberation and encouragement this important work can bestow.

Robert L. Powers
Jane Griffith

April, 2012
Port Townsend, Washington

FOREWORD TO THE FIRST EDITION

Whoever has grasped and understood the concept of the unity of the personality will know that he must treat the individual's style of life, and not his symptoms.

Adler (1963, p. viii)

This text is a comprehensive study of a technique and method for personality assessment. It is designed to enable clinicians to understand individual uniqueness and to teach and inform them how to convey this understanding to their clients or patients.

Following a thorough in-depth explanation of the theoretical concepts of Alfred Adler's Individual Psychology, the authors apply these concepts to the practice of psychotherapy, specifically to the task of understanding the individual style of life.

Detailed case histories, enriched by frequent verbatim dialogues, are used to guide the reader through the assessment process. These histories are most elucidating to the reader, even as they were to the individuals whose cases were being discussed. In each instance the lifestyle of the person emerges with full clarity.

Moreover, the reports are instructive for demonstrating how concepts are actually applied in therapy, as well as how to use the process the authors present in the text.

For a full understanding of a person's lifestyle and personality that is consistent with the theoretical concepts of Individual Psychology, this book is of extreme value.

Kurt A. Adler, M.D., Ph.D.

March, 1987
New York, New York

FOREWORD TO THE SECOND EDITION

The publication of the second edition of this book is very welcome. Adler, though sometimes overlooked, is increasingly acknowledged for the brilliance of his theory and its practical applications that have informed many other therapies developed over the years.

Adler's work, now validated by neuropsychology, emphasizes humans' inborn capacity for social connection, and shows how we creatively interpret experience to achieve belonging and significance in the wider world. Adler saw that uncovering and understanding an individual's unique movement is essential to successful engagement in psychotherapy.

Bob Powers and Jane Griffith are noted Adlerians with an unparalleled understanding of Adler's theory and a wealth of experience in its practical applications. Their book, therefore, has a richness seldom achieved in psychotherapeutic writing.

I have taught Adlerian Psychotherapy for many years and found the first edition invaluable as a teaching text. It successfully combines theory with detailed, exceptionally well-designed case examples for teaching the practical applications of the approach.

The new edition includes numerous exercises and illustrations, and brings fresh attention to ethnic, racial, and sexual diversity. It will be a welcome resource for teachers and students of Adlerian Psychotherapy, as well as others in the field.

Robert Armstrong, Ph.D.

April, 2012
Vancouver, British Columbia

ILLUSTRATIONS

THE KEY TO PSYCHOTHERAPY
UNDERSTANDING THE SELF-CREATED INDIVIDUAL

CONTENTS

A SPECIAL NOTE

PART I: ORIENTATION

The uncovering of the . . . style of life for the patient is the most important component in therapy.

Alfred Adler (p. 334)

Chapter 1. THE UNIQUE STYLE OF LIVING

This book comes out of the history of Alfred Adler's Individual Psychology, sometimes regarded as a division or subsection of "depth psychology." The history of depth psychology, in turn, has its place against a background of what Henri F. Ellenberger (1970) called *The Discovery of the Unconscious: The History and Evolution of Dynamic Psychiatry* in the title of his monumental study of the subject.

The modern chapter of this history lay in nineteenth-century medical experiments with hypnosis, which supported the hypothesis of an unconscious dimension of human experience and thought, limiting, perhaps dictating, the forms of conscious processes. This idea became central in the search for a "cure" for a variety of peculiar patterns of behavior, then coming to be understood as "illnesses," even though examination of their sufferers did not yield relevant findings of organic disease.

Adler, Sigmund Freud, and Carl Jung were pioneers in the effort to carry medical diagnosis and treatment into the assumed depth of this unconscious dimension. The spatial metaphor of depth was meant to suggest a hidden spiritual (psychic) reality lying not only beneath the surface of observable behavior, but also beneath even our subjective, or conscious, experience of ourselves. If that hidden reality could be uncovered, discerned, and understood, it might yield the reason for (perhaps even the cause of) all the irrational, bizarre, and otherwise inexplicable actions human beings occasionally exhibit. Even more intriguing was the thought that this hidden realm might be entered through some access into its secret dynamic, and that correctives could be applied to its disorders.

The language at hand for discussing the scope of this pioneering task was that of western religious and philosophical thought, a language of soul and body, mind and matter. It carried with it a variety of associations, Biblical, Platonic, Aristotelian, Newtonian,

3

and Cartesian. Adapted to the frame of reference of the new discipline and profession of scientific medicine, especially of its newest branch, psychiatry, in which the cure of souls (Gr., *psyche*) was becoming a province of the physician (Gr., *iatros*), it was a language rich with connotation, and therefore imprecise.

As pioneers, Adler, Freud, and Jung had also to incorporate the new Darwinian assumptions of a continuity between "nature" (already an ambiguous term) and "human nature" (an even more ambiguous one), and to do this against the dualism implied by much of their language. Further (and the importance of this is still far too little appreciated), they were working in a time of mounting democratic critiques of venerable institutions and hierarchical structures hitherto revered as inherent in the requirements of social order. Traditional ideas about the individual's place in society were being called into radical question, especially ideas of "high" and "low," "superior" and "inferior."

They took on an enormous task, challenging all their considerable powers of intellect and creativity, appealing to their not inconsiderable personal ambitions, and drawing upon their often unexamined values and biases. It cannot be any wonder that they came to disagree.

Doctors Disagree

The major disagreements arose first between Adler and Freud; soon after, between Freud and Jung. Adler's writings do not record his disagreements with Jung's Analytic Psychology in any substantial detail. In his later writings Jung recorded his respect for Adler's work (while remaining critical of certain elements in it), and included a recommendation to his own students that they also study Adler.

The disagreements that interest us here, therefore, are those that ended the collaboration between Adler and Freud, and that led to the creation of Adler's Individual Psychology as a theoretical system distinct from Freud's Psychoanalysis.

In a rather rough abstraction from the ways in which their lives and their careers connected, impinged on one another, and diverged, we can locate these disagreements in three major areas: the field of inquiry; the framework for understanding; and the purpose, desired outcome, and method of corrective intervention.

Field of inquiry

For Freud the field of inquiry was physiological, instinctual, and internal. He wanted to identify universal structural determinants that could be seen as governing personal development in any society.

For Adler it was biological, cultural, and transactional. He was more interested in the contexts of varying historical circumstances that could be seen as limiting personal participation in social development.

Framework for understanding

Freud postulated instinctual forces that drive the person into inescapable conflict with the requirements of social living. He saw this conflict as experienced internally in tension between hungers for gratification and needs for security.

Adler assumed that the person, having evolved in the context and matrix of social living, which provides the distinctive ecological niche for human adaptation, is suited only to the life that social living requires.

Adler recognized that social requirements, being symbolically transmitted and apprehended, are susceptible to misunderstandings and errors, both in their transmission and in their apprehension. He was unambiguous in arguing that the forms of social living, being subject to cultural variation and evolution, also require further development. To be well adapted to the requirements of social life is also, consequently, to participate in this social development.

In spite of this, critics of Adler persist in reducing his description of the requirements for successful social adaptation to the level of a shallow prescription of conformity. These critics were and they remain unwilling, perhaps unable, to appreciate a perception of individual and society in ecological terms. (Seeing the relationship between individual and society in adversarial terms does not, however, seem philosophically questionable to them.)

Purpose, desired outcome, and method of corrective intervention
For Freud, troubled persons must gain the strength to accept a tragic demand, namely to sacrifice the largest part of gratification to security. The desired outcome of therapy is insight and resignation. The method of intervention involves a "transference" onto the physician of the patient's conflicting desires for gratification on the one hand and "dependency needs" on the other, all of which were originally focused on the parents. The conflicts are then "worked through" in the transference.

For Adler, troubled persons must gain clarity in their awareness of being participants, equally active with others in shaping the community to which they belong. The desired outcome is a liberated sense of engagement. The method of intervention provides an experience of collaboration with the physician, in which the patient discovers and rehearses what parents and other educators had failed adequately to cultivate, namely a clearer sense of choosing (therefore sometimes refusing) equally with others among the many opportunities for making useful contributions, and consequently for enjoying the satisfactions of effectiveness and participation.

In line with his positions in these three areas, Adler was suspicious of the metaphor of depth, which he used sparingly and ironically, and even then chiefly as a reference to *context* (G., *Zusammenhang*). By this he meant first, the particular context of the individual's unique style of living that provides the indivisible pattern of consistency connecting the person's every movement; and second, the larger context of the social or ecological situation in which individual movement is indivisibly embedded. In this ecological emphasis he acknowledged his affinity to the Gestalt psychologists and their experiments in the perceptions of figure and ground.

The Assessment of Uniqueness

The ability to assess and understand an individual's unique style of living was therefore of central importance to Adler. It retained this importance for those who originally studied with him, and it is still central to those who are building upon and continuing to develop the implications and applications of his approach.

The question was always, how is this assessment to be done? Those who witnessed or experienced Adler's work, and whose reports we have, all included some astonishment in their accounts of it. Rudolf Dreikurs, a major proponent of Adler's work, in an interview filmed long after the event (Allen, 1969), still sounded breathless as he recalled his skepticism on hearing Adler claim that all the information necessary for understanding a person could be obtained in one hour. Many of Dreikurs's students can recall, in a similar way, their first experience of seeing Dreikurs prove that it might sometimes be learned in less than an hour.

Nevertheless, assessment of uniqueness was and still remains a difficult thing to teach. Adler described Individual Psychology as a "science for the understanding of persons" (in German, *Menschenkenntnis*). Because students find it so difficult to see how this understanding can identify the pattern of a person's movement, they may refer to its practice as "magic" upon first encountering it.

To consider a parallel, there is no one way to score the pattern of movement in dance. Although choreographers may employ their own systems of notation, none of these systems approaches the universality of a language, as in a language of musical notation. A musician reads a score, as an actor reads a script. In contrast to this, the dancer *follows* the movement of the choreographer. By analogy to psychotherapy, the client is at first the choreographer. The therapist is the dancer, following the pattern of movement, and sensing its logic and rhythm. As this is taking place their change of roles has already begun, as the client learns, even from the therapist's following the logic of the steps, how patterns of personal movement can come into harmony with larger circles of social movement.

But of course movement in the individual's unique style of living is not art for art's sake. Rather, "Human life . . . expresses itself in movement and direction toward a successful solution" (Adler, p. 163). The movement of a life is in the direction of a personally defined goal of success; any step in that movement is intelligible only as it is seen as serving to maintain the individual's line of direction toward that goal.

In an assessment of uniqueness one must therefore be prepared to see that *everything is in anything* in a holistic rather than in a linear apprehension. It defies the ordinary uses of language to speak of such a task. Since however we must speak of it in order to teach it, we will risk using artistic metaphor to avoid the greater risk of reducing the holistic to the linear.

We therefore sum up our difficulty by borrowing a classic exclamation and question from William Butler Yeats (Untermeyer, 1942, Part II, p. 128):

> O body swayed to music, O brightening glance,
> How can we know the dancer from the dance?

We must avoid thinking in terms of the lifestyle of a person apart from the person. A person does not *have* a lifestyle, in the sense of its being attached like a possession, a piece of clothing, or a shadow. It is closer to the sense of one's having an accent of speech, which may reveal something of the context in which one learned to talk. To understand lifestyle is to think in terms of movement and pattern. It is to recognize an indivisible person moving in a characteristic style in every situation, and to know how to identify every thought, feeling, and action of that person as having its place in coherence with this unmistakably unique pattern.

In fact, we perceive pattern and movement without being aware of it in our everyday identification of one another. Walking on the street we may recognize someone familiar to us long before we are able to distinguish the details of facial or other features. It may be someone whom we have not seen in years. Yet there is already something in gait, posture, and pace that we recognize. As we draw closer each newly revealed detail fits in the overall impression to confirm recognition: "Well, there's old Bob Taylor! I'd know him anywhere."

We also rely on our ability to perceive pattern and movement for a sense of predictability in our dealings with others. We are aware of certain probabilities as to how a person known to us will think, feel, and act in a given situation: "If George thinks you're loafing he'll be furious and he'll fire you in a minute." We are also aware of how lifestyle shapes responses to the social world, independent of reality: "She's the sort of woman who could start a fight in an empty house" (Irish saying).

The Law of Movement

What psychologists do in the assessment of the *individual variant* (Adler, pp. 179-180) is a refinement of the commonplace mode of perception and discrimination described above. Through the mastery of a set of skills and the practice of an imaginative empathy we seek to get the sense of an individual's law of movement. When we understand this, we can see that in this person's situation (as this person experiences it) any behavior that seemed puzzling or even alien until now is in fact (given the "private sense" of this one human being) intelligent, hence intelligible.

We define the key to an individual's law of movement by reference to belief and attitude, words connoting evaluation and predisposition. We want to evoke the image of a person's basic preparation for social living laid down in infancy, before language was available to express thought, before feelings were rehearsed out of sensation, and before action was independent of the random movements of trial and error. In that time before *a* time could be distinguished, we were already in a world of social being, already beginning to sense the extent of our being welcome or unwelcome in it in our striving. We were learning how to move in it wherever we had or could enlist the assent and support of others. We were alert to the extent to which we were unimpeded by opposition or reproach. We were moving within a genetically given range of possibilities, including a greater or lesser degree of activity that marked our efforts to make a place for ourselves among the others. We had a sense of what was possible, what was invited, what we might do to proceed. "Once upon a time," *in illo tempore* (Eliade, 1959, p. 21), we were sizing up our world, ourselves, and our prospects. We were beginning to shape our movement in harmony with evaluations we would later argue for in language and support by feelings. Our attitude (to borrow an image from sculpture) was displayed in our posture, that is, our readiness to move — toward, away from, or against others.

In this movement of our early childhood each of us formed the prototype of a unique style of living. We were already inclined (another postural image) to see things in a particular way, and to appraise our opportunities on the basis of our expectations of what would unfold. As we developed out of infancy our experiences were more and more subject to these expectations, and our perceptions were more and more selected and shaped to conform with them. We were reaching the point where a particular impression of a single event could be chosen to illustrate and confirm the validity of the general principles on which we were basing our efforts. Therefore Adler, in his examination of a patient, looked to the person's earliest recollections from childhood to locate the first notes of a melody that is by now fully developed and well-rehearsed, in which a person continues to express an attitude toward all the experience of life. Adler (1964) attended to what was presented, that is, the practiced tune (and tone), not only in its parts, but in the whole, and in its resonance of meanings: "The style of life commands all forms of expression; the whole commands the parts" (pp. 174-175).

Our way of persisting in these meanings, in efforts to make sense of the world, is revealed most clearly by exaggeration, and recognized most unmistakably in laughter. Not long ago the *New York Times* ran a photograph showing a man standing beside his expensive new car in an upper level of a parking garage. The rear end of the car was tilted up; the font end had broken a protective barrier, and had gone through an open space to rest on the roof of an adjoining structure just below. "My whole life has been like this," the man was quoted as saying. Commonsense tells us that his whole life could not have been like this extraordinary incident, but his belief is independent of commonsense. Understanding lifestyle depends on an ability to recognize the personal symbolism and attendant meanings that a person may believe are illustrated in a single, and sometimes exceptional, event.

The Unique Orientation

A child gives meaning to circumstances, and the meaning of life that is first a child's creation serves as the starting point for a system of *orientation*. Such a system is the primary, inescapable, and essential basis for progress through the human world of symbols, values, and the uncertainties of choice, risk, and experimentation. Being the work of a child, it was at first narrowly conceived and only partly subject to the corrections of a larger world of social experience. "Truth will sooner come out from error than from confusion," said Francis Bacon (1620, *Aphorism 20, Book II*) at the dawn of the modern era. He was alluding to the intellectual situation of the experimental scientist; he could just as well have been reflecting on the existential situation of every child.

Adler (p. 376) recognized that the situation of any child was inherently different from the situation of any other: "It is a common fallacy to imagine that children of the same family are formed in the same environment." Not only are no two children born into the same environment, but no two are born into the same family. Each child coming into the world alters it by entering. It is not enough to ask how or to what extent parents and siblings share an influence on the child's development. By virtue of being born into the family a child has already changed the lives of both its parents and siblings. The firstborn changes the social status of the parents, ordaining husband and wife to the irrevocable social roles of mother and father. The secondborn dethrones the first from the eminence of onliness, and enters not only as a child but as one of the children (perhaps, if there are no others, as the one to be known as "the baby" forever after). And so on. Even in multiple births there is an ordinal position, and a meaning attached to it. (In the course of a public lecture, a famous physician and author reflected on her position as the secondborn of triplets, and on her experience of herself as squeezed between her older sister's claim to a position of prerogative and her younger sister's hold on a position requiring indulgence.)

The law of movement is therefore applicable only to the single case, that is, to the individual variant, the unique creation of an individual who works it out in a situation that is unlike any other. Where the situation supports efforts to move toward success, offers opportunities for participation, and conveys respect for the ways in which the child is able to make contributions, we expect the person's law of movement to tend toward the useful side of life in response to these encouragements. The child is likely to give a meaning to life that is generally useful in that it enables cooperative and respectful movement among others, and so provides the basis for an effective and satisfying adult development. To the extent that a person's basic convictions reflect this meaning, there is likely to be little or no reason to inquire into them, to examine their content, or to ask how or for what purpose they were formed.

Even in the case of such an encouraged person, an unexpected loss, or betrayal, may present itself, requiring greater courage and a greater capacity for cooperation than the person's law of movement has trained for. Lifestyle assessment and psychotherapy may be useful for the recovery of equilibrium in such a crisis. Otherwise, in the absence of the shock of such a challenge to the style, people are free to go through life with biases intact, and with the happy ability to make friends, make a living, and make love.

Sad to say, it must also be said that there are many childhood situations that can be seen only as "overburdened" (Adler, p. 367). A crushing sense of organ weakness or other disadvantage may be unrelieved by the presence of an ally, or by an inspiration to see what could be done to overcome it. A pattern of pampering may limit a child's sense of competence to a hazy awareness of how victories are gained by displays of suffering, weakness, accusation, or complaint. Any form of parental hatred or neglect may thwart efforts directed toward success in socially useful behavior, leaving only a negative sense of how triumphs can be won through maneuvers of conflict, revenge, or avoidance. We cannot expect a child's law of movement, if it was devised and rehearsed in adaptation to situations such as these, to rest on meanings other than those that are mostly safeguarding, private, and useless for a socially effective and satisfying life.

The basic discouragement of such a beginning enters into the person's expectations, partly directing development toward further experiences of discouragement. A dread of further loss constricts the capacity to share the commonsense of participation in the give and take of life, and fosters a private sense in its place. Possible difficulties appear as threatened defeats in a lifestyle that deviates more and more, by means of a variety of cautionary warnings and arguments, from the promise, the value, or even the possibility of cooperation and contribution.

Whatever probabilities may suggest themselves, it is important to resist an intimation of determinism. Any individual who comes to a counselor out of an overburdened and discouragenic background draws upon remnants of courage retained in hope (even if the hope is often veiled by expressions of suspicion and disdain). Such a move proceeds on the basis of some glimmer of an awareness, however reluctant, of one's having created and maintained one's own law of movement with a meaning that will not work.

There are myriad variations in the ways different individuals prepare themselves for life. Each of these is bound to incorporate some element of error. As children, none of us could have devised an encompassing and balanced view of life's meaning, or could have prepared ourselves for any and every contingency we might encounter in meeting life's challenges. A discrepancy between our practiced style of living and some demand for a creative solution to a novel and unexpected problem will present a crisis. A "good" childhood cannot prevent that; a "bad" childhood does not produce it. Whenever such a discrepancy occurs it reveals a failure to understand the extent to which all the problems of life, being social problems, require cooperation for their solution. Whether it is in a sudden shock of betrayed confidence, or the final culmination of despair over a chronic pattern of misfittedness, life does not yield; it is the lifestyle that is called into question.

The problem remains: How are we to derive an understanding of the unique style, and how are we to convey a useful awareness of its distortions to the person who resorts to us for the task of correcting errors and reconsidering goals through the processes of psychotherapy or other counseling? How, to quote "an English author" (to whom Adler referred, but did not name), can we learn "to see with the eyes of another, to hear with the ears of another, to feel with the heart of another" (p. 135)?   This book shows a way to answer these questions.

A SUMMARY OF
INDIVIDUAL (ADLERIAN) PSYCHOLOGY
THEORY

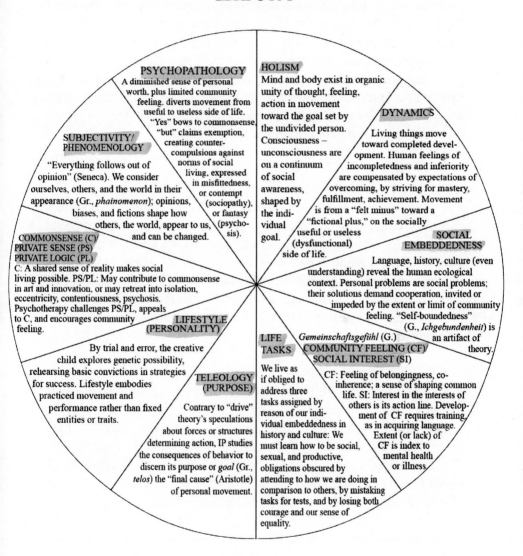

PSYCHOPATHOLOGY
A diminished sense of personal worth, plus limited community feeling, diverts movement from useful to useless side of life. "Yes" bows to commonsense, "but" claims exemption, creating counter-compulsions against norms of social living, expressed in misfittedness, or contempt (sociopathy), or fantasy (psychosis).

SUBJECTIVITY/ PHENOMENOLOGY
"Everything follows out of opinion" (Seneca). We consider ourselves, others, and the world in their appearance (Gr., *phainomenon*); opinions, biases, and fictions shape how others, the world, appear to us, and can be changed.

COMMONSENSE (C)/ PRIVATE SENSE (PS)/ PRIVATE LOGIC (PL)
C: A shared sense of reality makes social living possible. PS/PL: May contribute to commonsense in art and innovation, or may retreat into isolation, eccentricity, contentiousness, psychosis. Psychotherapy challenges PS/PL, appeals to C, and encourages community feeling.

LIFESTYLE (PERSONALITY)
By trial and error, the creative child explores genetic possibility, rehearsing basic convictions in strategies for success. Lifestyle embodies practiced movement and performance rather than fixed entities or traits.

TELEOLOGY (PURPOSE)
Contrary to "drive" theory's speculations about forces or structures determining action, IP studies the consequences of behavior to discern its purpose or *goal* (Gr., *telos*) the "final cause" (Aristotle) of personal movement.

HOLISM
Mind and body exist in organic unity of thought, feeling, action in movement toward the goal set by the undivided person. Consciousness – unconsciousness are on a continuum of social awareness, shaped by the individual goal.

DYNAMICS
Living things move toward completed development. Human feelings of incompletedness and inferiority are compensated by expectations of overcoming, by striving for mastery, fulfillment, achievement. Movement is from a "felt minus" toward a "fictional plus," on the socially useful or useless (dysfunctional) side of life.

SOCIAL EMBEDDEDNESS
Language, history, culture (even understanding) reveal the human ecological context. Personal problems are social problems; their solutions demand cooperation, invited or impeded by the extent or limit of community feeling. "Self-boundedness" (G., *Ichgebundenheit*) is an artifact of theory.

LIFE TASKS
We live as if obliged to address three tasks assigned by reason of our individual embeddedness in history and culture: We must learn how to be social, sexual, and productive, obligations obscured by attending to how we are doing in comparison to others, by mistaking tasks for tests, and by losing both courage and our sense of equality.

Gemeinschaftsgefühl (G.) **COMMUNITY FEELING (CF)/ SOCIAL INTEREST (SI)**
CF: Feeling of belongingness, co-inherence; a sense of shaping common life. SI: Interest in the interests of others is its action line. Development of CF requires training, as in acquiring language. Extent (or lack) of CF is index to mental health or illness.

Figure 1.

Finally, the present book is designed to address a specific application of the theory and practice of Alfred Adler's psychology; it is not intended to be a general text in Individual Psychology. Therefore, for those who have not yet examined this approach in previous study, we conclude this chapter with a summary representation of the theory in Figures 1 and 2. Use of a pie chart in Figure 1 is meant to suggest the coherence of the theory's elements as holistic and interconnected, rather than analytic and linear. Figure 2 represents Individual Psychology's understanding of personality dynamics.

Conclusion

In this chapter we have presented a brief review of Adler's concepts concerning lifestyle development, and an exposition to the ways in which the creative child gains an orientation in the world. At the same time, we have offered an orientation to the reader by which to approach the task of assessing lifestyle with an attitude of discovery and enhanced respect for what the client has achieved out of the numberless possibilities open in the childhood situation.

In the first part of Chapter 2 we discuss the process we employ in the assessment of a lifestyle with the help of an illustrative case example. In the second part of Chapter 2 we review various instruments of assessment suggested by Adler and others. We then introduce the format which the present text shares in outline with *The Individual Psychology Client Workbook (IPCW)* (Powers & Griffith, 2011), published as a separate volume to be used in clinical practice as a guide to lifestyle assessment.

PERSONALITY DYNAMICS
The schematic abstracts extremes for teaching purposes.

All striving is directed toward overcoming universally experienced feelings of inferiority and incompleteness to achieve a subjectively conceived, fictionally defined, goal of success: *the personality ideal* (*Persönlichkeitsideal,*G.). Striving toward the goal takes different forms, but the goal remains.

THE GOAL OF SUCCESS
The Fictional "Plus" Position

The Useful Side of Life	The Useless Side of Life
Contribution, working to the advantage of all	Self-promotion, working at the expense of others
Goal of personal mastery and fulfillment, in line with the needs of the situation	Goal of personal superiority over others, in disregard of the needs of the situation
Striving of the encouraged person, who feels a strong sense of belongingness	Striving of the discouraged person, who feels a diminished sense of belongingness

INFERIORITY FEELINGS
The Felt "Minus" Position

Figure 2.

Chapter 2. PSYCHOCLARITY AND THE PSYCHOCLARITY PROCESS

In 1982 we coined a new word: psychoclarity.

Neologisms appear whenever familiar terms are no longer equal to the tasks we set for them. The familiar term in this case, psychotherapy, is a word we do not expect our new word to replace. Our hope is, rather, to enhance its meaning and to support a clearer use of it.

According to the *Oxford English Dictionary*, the term psychotherapy is itself a neologism, first used in 1904 as a simpler form for psychotherapeutics (meaning, in its original usage, the medical treatment of physical illnesses by psychological means). Subsequently psychotherapy came to refer to the process in which one person seeks the guidance of another in quest of relief from any of a variety of painful, often troublesome, feelings, thoughts, or actions.

The usefulness of this term remains limited, however, by its association to medical procedures, at least where these continue to be understood as *directed* by active persons (doctors) *toward* and *upon* passive ("patient") persons. Many physicians have come to regard that model as deficient, but its implications continue to obscure an understanding of what is entailed in psychotherapy. Use of evocative but undefined medical metaphors of health, emotional or mental, also cloud a practical guiding image of the desired outcome of the process.

It therefore seems to us legitimate to reach for a new term to refer to the quality and form of the collaborative relationship between client and clinician in psychological counseling.

When we first put the term psychoclarity forward (Powers & Griffith, 1982), we wrote (somewhat amended here):

18

> The client is invited to help the therapist to see more clearly what is in the client's thought, feeling, and action, and what its effects are upon the social field. . . . Client and therapist are united in the effort to understand what is going on, what it goes on *for*, and what it costs to keep it going. . . . As the therapist comes to see more clearly, the client does as well.

We cautioned against the transaction typical of the more traditional approaches in therapy:

> Many therapists are trained as if their task were to pursue the goal of changing the client. . . . The therapist who wants the client to change is setting up for a fall. Only the client can decide when to change and what and how to change. The psychoclarity formula is: You cannot change your mind until you know your mind. You cannot know your mind until you can speak your mind, and your speech brings the private [sense] into the commonsense world. When that occurs, your private sense of things is no longer *unspeakable* and therefore, no longer *unthinkable*.

Therapy gives the client an opportunity to break the spell of the unspeakable by clarifying for the therapist his or her private sense of what it is to be in the world. When I know my mind I can see that while the past behavior is now *understandable*, it is no longer *necessary*.

The goal is one of *self-understanding*, without which a consideration of whatever might be changed is impossible. Therefore we went on to say:

> Only at this point is the client free to reconsider, and in that liberation may decide to do so.

Psychoclarity is our word. However, our emphasis on collaboration, self-understanding, and psychic liberation must be attributed to Adler (1980), who stated it this way in 1931:

Treatment itself is an exercise in cooperation and a test of cooperation. . . . [The client] must contribute his part to our common understanding. We must work out his attitudes and his difficulties together. . . . We must cooperate with him in finding his mistakes, both for his own benefit and the welfare of others.

The usual treatment has been to attack the symptom. To this attitude Individual Psychology is entirely opposed. . . . What we must always look for is the purpose for which the symptom is adopted and the coherence of this purpose with the general goal of superiority. . . . With the change of the goal, the mental habits and attitudes will also change. [The client] will no longer need the old habits and attitudes, and new ones, fitted to his new goal, will take their place (pp. 72, 62-64).

"The Arrangement" in the case report that follows illustrates these concepts. The case is further illustrative of the imagery presented in Figure 3 (p. 21), distinguishing between the horizontal and vertical planes of movement, proposed by the original work of Lydia Sicher (Griffith & Powers, 2007, p. 56). At the conclusion of our presentation of this case we will return to a discussion of the psychoclarity process.

The Arrangement

A psychologist asked a fearful and hesitant, but intellectually vain, client, "Do you know the proverb, 'Nothing ventured, nothing gained'?" "Of course," the client replied. "Can you tell me what it means?" the therapist went on. "It's simple," the client explained patiently. "It means that if you don't try anything you won't fail." There was not a hint of guile or irony in his tone or his expression. He was innocent of how he had turned things around.

It was not a matter of organic damage, or a failure to understand the words. It was certainly not an intellectual inability to grasp the

HORIZONTAL vs. VERTICAL MOVEMENT

Movement on the Horizontal Plane

— Individual focuses on the goal of personal evolution toward mastery, competence, and fulfillment, in harmony with the continuing evolution of the community;

— expresses attitudes of equality and respect toward others;

— has a task-centered/problem-solving orientation, asking, "What does the situation require?" with the conviction, "We can work it out."

— Individual evaluates the difficulties of life as problems to be addressed in cooperation with others;

— directs choices toward the useful side of life, where allies are engaged in common concerns;

— expresses a developed community feeling, compensating for feelings of inferiority;

— is courageous, pressing on in the face of disappointments or set-backs.

Movement on the Vertical Plane

— Individual focuses on the goal of self-elevation at the expense of the community;

— presents a posture of being above others, compensating for fears of falling short;

— has a prestige/status orientation, asking, "How am I doing?" with the pre-occupation, "How do I look?"

— Individual treats the ordinary difficulties of life as personal affronts; hypersensitive to opposition or criticism;

— busy on the useless side of life with matters not of common concern;

— expresses diminished community feeling; competitive, tends to regard others as adversaries;

— limits courage to maintaining pretenses, as if in danger of "put-downs."

Figure 3.

idea expressed by the proverb. It was, rather, a *use* of brain, understanding, and ability in the service of a basic and faulty attitude. What the client (whom we shall call Edward) said was true in a sense; the proverb allows for the negative implication he derived from it. *In a sense.* That is the key phrase. The sense he made of it is not the sense we commonly take from it.

Commonsense is a sense we share with others, a feeling for the way life is shaped, for what it requires of us, and for what we have to offer in meeting its requirements. Edward's *private sense* led him into a private way of thinking that could not be shared; it was the basis for a *private logic* that could not be followed. In fact his private sense was opposed to the sense commonly expressed by use of the proverb, arguing against the courage required by life, and withholding assent from the willingness to take the risks and accept the set-backs entailed in any pursuit of success. The safeguards his private sense raised against his exaggerated opinion of the danger of defeat had become barriers against the demand to take part. His private argument concluded that the struggle to solve problems and overcome difficulty must be left to others. Finally (and for him, the worst of it), his sense of life did not work, and he was suffering.

Edward's safeguard against failure had failed. It could not protect him from being passed over for a desired promotion, it did not allow him to enjoy the company of friends, and it left him sexually impotent.

His wife was an affectionate, capable woman, who was advancing in her career. When he pointed to the jealousies of others and their failures to appreciate his brilliance as the reason for his difficulties, she was sympathetic. That didn't help. He only suspected that she must, secretly, be looking down on him. His safeguards did not protect him from this suspicion, or from the distance he began to feel between himself and his loving ally.

He did not know what was wrong. He could not, because he could not think clearly about it. He could not think clearly about it, because he could not put it clearly into words. To the extent that he could talk or think about it at all, he missed the point, complaining about the spite, insensitivity, or jealousy of others. He could talk about these things, at least to his wife, and so he could think about them.

He could not talk about his basic convictions and attitudes, least of all with his wife, whose successes reminded him of his failure. To put such things into words would have revealed the extent to which they were antagonistic against everyone, even against her. Had he stated his convictions, their defiance of commonsense would have been revealed and his solitary rebellion against social living would have been exposed, even to him.

He could not say that he must dominate others to feel safe, that he must look down on others (including her) to feel worthwhile, or that he must impress her with his masculinity before he could respond sexually.

Language is a vehicle for meaning. *Meaning is social. The meaning of meaning is that it can be shared.*   Edward could express his attitudes only indirectly therefore, and in disguise. He cloaked his ambition toward domination by unarguable statements about not letting others "push me around" or "force me to compromise my ideals." In the same way, he disguised his posture of looking down on others by focusing on shortcomings which anyone might agree were real. For example, he kept his wife off balance by reminding her of inconsequential faults about which she was embarrassed, always with the "good intention" of helping her to correct them.   To account for the withdrawal of his sexual feeling, he criticized her for her "aggressiveness" in approaching him sexually, and told her that she was losing her femininity.

In a sense (private sense) he believed these things, even if uneasily. There was nothing else he could do, because there was nothing else he could say.

Thought is framed in language, which is a social possession; one thinks in a language which is not one's own. Here, as with the meaning of language, rational thought is thought that can be shared. It need not be, of course. We can keep thoughts to ourselves, as every discreet person knows. We can also state them, if we choose to speak our mind. Only the unspeakable remains unthinkable. *We can think clearly only in thoughts that can be stated plausibly to others.*

Edward was unable to state his basic convictions and attitudes because they did not make sense (commonsense). As a result, they could not take form even in his own mind as rational. He had to make do with his private sense, muddled in the painful awareness that something was wrong.

Instead of a clear awareness of himself as a participant in the common life, he had constructed an "arrangement" (Adler, p. 263) by which to picture his painful situation to himself. He had to do this in a way that did not add to his feeling of falling short and of having a status of diminished worth in comparison to others. His *attitude* was that of a man who would be above others; his *conviction* was that otherwise he would be nothing at all.

His strategy, which he was unable to see, was to deflect his feelings of diminished value off onto others; first, onto his wife, the person most disposed to be his ally. He also drew on her willingness to do her part and to make allowances for his insulting behavior, as if he were entitled to this resource to shore up his faltering self-esteem. This was a dangerous game, comparable, as Adler put it, to writing checks against the accounts of others. It was also self-defeating, because the more she struggled to understand and help, the more she appeared to him as the "better" person, adding further to his humiliation and leaving him with no way to deal with it other than by his strategy of counter-humiliation, which he intensified. Finally, she threatened to leave him. Edward's arrangement had begun to fall apart, and the remnants of his commonsense brought him to consult a psychologist. He was thirty-one years old.

Edward was a firstborn son. At the time of our first meeting his father was sixty-four years old, still an active research physician. His mother was sixty, an excellent cook and hostess, more modestly educated than his father, but polished in her speech and manners. She adored her husband, looked up to him as if he were a god, and did all she could to make life comfortable for him. Father had been kindly, not punitive, but his busy professional life left little time for the boy. He noticed Edward's accomplishments, and was especially pleased by signs of intellectual ability. Father did not enter into conflicts with his colleagues, among whom his position was one of unquestioned respect. He was, however, disdainful of mediocrity (as he often referred to it) in his peers or among his students. This disdain was never openly directed toward his son; rather, it was expressed, as the boy saw it, by a lack of interest in anything unrelated to academic excellence. His younger sister seemed to have an easier time with Father. Father was playful with her, and lavished approval on her appearance, the way she decorated her room, and the help she gave Mother. She did not do as well in school as Edward did, but Father never seemed to be bothered by that.

It was not hard to see, at least in rough outline, how Edward had formed his convictions, or how he had practiced his attitude. He was trained and had trained himself for a position above others. Superior achievements were expected of a man; anything less would be mediocre, and to be mediocre would be the same as to be nothing at all.

It was hard for Edward to see it. It sounded simplistic and stupid (as he put it), and he was brilliant. It was harder for him to see that his ambition was to be worth more than others, or to see how this ambition required others to be worth less than he, and to admire his superiority. It was hardest for him to see that his attitude and convictions forbade him any experience of fellow feeling or ease in friendship; that they spoiled any satisfaction in doing a job unless he was outdoing others; or that they robbed him of the refreshment of closeness and comfort in love, since they required that his partner remain at a subordinate distance.

He began to recognize what he had been trying to construct in his arrangement when the psychologist noted that he was then, at the age of thirty-one, approaching the same age his father was when he, Edward, was born. He was startled by this observation. He said he had been puzzled by an uneasiness at the idea of his next birthday, a feeling he had never had before. His startled reaction and his awareness of uneasiness were intellectually intriguing to him, suggesting that some unconscious (that is, "not-understood," Adler, p. 192) process might be at the root of his difficulties.

We had found a point at which to begin the work of psychotherapy; he had entered upon the process of psychoclarity. His meanings, for his age, his masculine guiding line, his expectations for himself and for life, all came clear in the subsequent assessment of his lifestyle. He was able to understand and to reconsider his convictions and attitudes.

Precursors

Structure reduces uncertainty; it strengthens the possibility of a useful process and beneficial results. Beginning with Adler, and continuing to the present text, Adlerians have developed a variety of ways to organize the elements involved in assessing an individual's unique style of living.

An *Individual Psychology questionnaire, for the understanding and treatment of problem children, formulated with explanatory comments by the International Society for Individual Psychology* was noted by Heinz L. and Rowena R. Ansbacher, premier editors of Adler's works, as being contained in several of Adler's publications, beginning in 1929. The Ansbachers report that Adler recommended this device as "an informal aid, not to be adhered to rigidly, in assessing a child's style of life, in learning what 'influences were at work when the child was forming it,' and in seeing how this style of life manifests itself in coping with the demands of life" (Adler, pp. 404-408).

Following this questionnaire is another headed, *For Use With Adults.* In an introductory note Adler (1964) is quoted as saying:

> In case of adult failures I have found the following interview schedule to be valuable. By adhering to it the experienced therapist will gain an extensive insight into the style of life of the individual already within about *half an hour* [italics added] (pp. 408-409).

A list of only eighteen questions is then offered. They follow the form of the usual medical history, beginning with, "What are your complaints?" and going on to the situation in which symptoms were first noticed, the current situation, and the person's occupation. The next question concerns the "character, and . . . health" of the parents and, if they are no longer living, the causes of their deaths. Then comes a question regarding the relationship of the parents to the patient. Only then is there an inquiry into the siblings, the person's place among them, and other particularities of the person's childhood situation.

Rudolf Dreikurs (1973, pp. 88-90) created a longer questionnaire to be used after acquiring information concerning what he called "the subjective condition" of the presenting complaint, its onset and history, and the current social or "objective life situation" of the person. He then directed the inquiry to the person's place in childhood among the siblings; in so doing, Dreikurs gave much more attention than Adler did to the relationships among them. A lengthy self-assessment of the patient's childhood standing relative to the other children in the family in terms of behavior, character, appearance, and other traits, was elicited through a series of ratings. Only then did Dreikurs turn his attention to the parents, their ages and characters, and the relationship between them.

Dreikurs's students adapted and amended his work. A *Family Constellation Interview Guide* was published by Bernard H. Shulman (1973, p. 57). Shulman and Harold H. Mosak (1971) produced *The Life Style Inventory*, a printed outline on which to record data. Daniel Eckstein, *et al.* (1982) expanded this in a form of their own as a *Life Style Interview Guide*. Raymond N. Lowe and Betty Lowe (Appendices H and I, Leroy Baruth and Eckstein, 1981) followed the same outline with further modifications in two separate forms, an *Interview Guide for Establishing the Life Style*, and *Life Style Interpretation*.

Don Dinkmeyer, Jr. and Len Sperry (2000) reported on and discussed a further treatment in (Roy) *Kern's Lifestyle Scale* and its expanded version, *BASIS-A* (pp. 303-306). All of these follow Dreikurs in his departure from Adler's sequence, keeping the inquiry about siblings prior to that about parents and other adults.

Dreikurs stressed the importance of this priority. To him, sibling relationships were of greater consequence to the individual's creation and rehearsal of a unique style than was the example of the parents or their efforts to mold or influence the child's development. Dreikurs emphasized the other pole of parent-child transactions, possibly because it was so often overlooked or disregarded in quests for the purported *causes* of personality formation. Following Adler, Dreikurs opposed speculations about causes, which lead (as he saw) toward pessimism, fatalism, and disrespect for individuals. In one of his most widely read books, Dreikurs put it unequivocally: "Earlier than we realize children shape and mold themselves, their parents, and their environment" (1964, p. 32). We agree.

We have returned to Adler's priority in our structure for the initial interview. Our purpose is to restore Adler's emphasis on *context*, to draw attention to the usefulness of contextual thinking, and to demonstrate its importance to students and practitioners of Individual Psychology. (Dreikurs was of course also aware of the importance of context. His term "family atmosphere" was clearly designed to evoke it, as was his use of Adler's early image of "guiding lines" to refer to unconscious rehearsals for gender roles.) When we go on to investigate the childhood family situation, we follow Dreikurs.

The Psychoclarity Process

The psychoclarity process rests upon a thorough appreciation of the context in which the inevitable errors made by a child can be understood as having made a certain kind of sense. If, as Adler put it, "every neurotic is partly right" (p. 334), a humane effort at identification may let us see how, as such a child in such a situation, we could have been creative enough to come up with the same mistakes.

Once this clarity is achieved, the client who was that child is able to consider those errors with a sympathetic understanding and, perhaps for the first time, may experience a sense of optimism about what life now holds open.

Until this clarity is achieved, the error cannot be seen at all; it can only be experienced in a kind of *psychomuddle*, accompanied by a feeling of inevitability. Since the error lies in an evaluation of life's requirements, and in a corresponding attitude toward them, the error inheres in the person's way of apprehending everything. A way of seeing does not see itself. Just as those whose vision deviates from the norm cannot account for the discrepancy between what they see and what others report seeing, a person whose style deviates in erroneous convictions and attitudes may be aware of missing something and yet be unable to account for it. A faulty visual apparatus is one of the genetic possibilities of the organism; a faulty attitude is one of the developmental possibilities of the creative person.

The lifestyle assessment procedure calls for a thorough inquiry into the particularities of the situation in which this one person learned to rehearse and to make do with a particular pattern of errors. This procedure reflects the structure of practice in which two people work with the client (see p. 39), one responsible for gathering the material, both participating in a review and interpretation of the information with assistance of the client, and one responsible for summarizing its implications. It allows for about ten hours to complete the work (following an initial interview of two to three hours). Sometimes the assessment takes as long as fifteen hours depending on the complexity of the client's personal history, on how quickly the therapists are able to make sense of the data, and on whether the client is verbose or laconic. We do not expect anyone to imitate the procedures set forth in this text in all their particulars. Clinicians report that they adapt our structure readily to the requirements of a variety of circumstances.

In any case, this procedure takes much longer than either Adler or Dreikurs demonstrated was necessary for "the experienced therapist." It is true that an extensive understanding of a lifestyle can be gained in an hour or even in half an hour when the basic principles of assessment have been mastered and practiced; in Individual Psychology, and in the psychoclarity process, it is also true that no fixed boundary is drawn between assessment (diagnosis) and psychotherapy (treatment). The client is engaged in psychotherapy at each step in the assessment. We therefore feel free to take all the time necessary for a systematic exploration of the story of this person's life and an attentive reflection upon the meaning of its difficulties.

The *Individual Psychology Client Workbook (IPCW)* presents a format; the present text provides a method for using it. Neither of these can be more effective than its users, a large part of whose effectiveness resides in the discretion with which they employ this material. In considering a client's history, the clinician must bear in mind that a numerical or even logical order does not dictate a fixed sequence for gathering data.

We are not in search of a body of fact; our goal is to discern the pattern of an embodied life expressed in movement, the key to which lies in convictions and attitudes. Dreikurs made use of Henry Winthrop's (1959) term, factophilia, to warn against the activity of collecting more and more data without a sense of the context in which it could be intelligible. He appealed to the image of a mosaic, which may reveal its pattern even when only some of the tiles are in place. A lifestyle assessment is not a systematic collection of a heap of tiles; it is the imaginative reconstruction of a pattern which allows each tile of information to be fitted into the context of the whole. This context, furthermore, is a pattern of movement in line with what Adler recognized as the "great line of action" of the "whole of human life . . . from below to above, from minus to plus, from defeat to victory" (Adler, p. 255).

Elements of the Procedure

The Initial Interview. This procedure calls for an initial interview of two to three hours. In this book, the initial interview process is set forth in three chapters: Chapter 3, "The Initial Interview Inquiry"; Chapter 5, "Life Situation, Presenting Problem, and Life Tasks"; and Chapter 12, "Impressions Derived from the Initial Interview."

The Lifestyle Assessment. The lifestyle assessment follows and builds on the initial interview. Chapter 4 describes the lifestyle inquiry; Chapters 6 through 11 illustrate the process of reviewing and interpreting the material gathered in the lifestyle inquiry; Chapters 13 and 14 offer instruction on the development of the final summaries.

The Early Recollections. Both the gathering and interpretation of early recollections have their own techniques that are integral to the lifestyle assessment. The process of gathering early recollections is explored in Chapter 4, which includes a section on special problems. Chapter 11 provides a method for interpreting early recollections, and Chapter 14 provides outlines and formats for constructing the two early recollections summaries.

Key Features of the Child's Experience

There are features common to the landscape of childhood to which every child attends with a unique evaluation. We isolate seven such features, the evaluations of which are expressed in a characteristic pattern of movement, revealing the basic convictions of the lifestyle and forming the foundations of a lifelong orientation. These are:

1. Masculine and Feminine Guiding Lines and Role Models

2. Family Atmosphere

3. Family Values

4. Other Particularities (the ethnic/racial, religious, social, and economic situations)

5. Birthorder

6. Genetic Possibility

7. Environmental Opportunity

The outline below relates each of these features to its evaluation, indicated by a phrase following the slash (/). The first statement, for example, represents the child's identifying the "Masculine and Feminine Guiding Lines and Role Models" in parents and other men and women in its sphere, and then constructing a pattern of responses, designated here under the heading "Gender Role Preparation."

The effort to illustrate these ideas in a diagram (Figure 4, p. 35) risks misrepresenting as static or determined the person's lively creating of a unique pattern of movement. The work relies on your imagination to see through to the creativity and subjectivity involved on both sides of the dynamic reality as you study the suggested *givens* in a child's situation (represented by the shaded, curved arrows) and the meaning and value the child *gives* to them in response (represented by shorter, unshaded arrows).

1.   *Masculine and Feminine Guiding Lines and Role Models / Gender Role Preparation*

The child evaluates the meaning of gender in preparation for taking part as a gendered person, concluding, "This is the way men are and the way women are. This is what I can expect of members of my gender and the other gender. Therefore this is what is going to unfold for me (as a man or as a woman) . . . unless I do something about it."

2. *Family Atmosphere / Interpersonal Style*

The child evaluates the family atmosphere, set by the attitudinal and operational relationship between the parents, and between the parents and the world (including the children), and concludes, "This is what I have to expect and prepare for in my dealings with others."

3. *Family Values / Personal Code*

The child considers the values shared by the parents, and concludes, "These are the issues of central importance, on which I must be prepared to take a stand." (Those values adhered to independently by either parent are not to be considered as "family values," but as features of the gender guiding lines.)

4. *Other Particularities (Ethnic/Racial, Religious, Social, Economic) / Limits of Community Feeling/Social Interest*

The child evaluates the ethnic, religious, social, economic, and other particularities of the situation, and concludes, "My starting place is defined within these particularities, and in my movement toward participation in the broader community, I carry my sense of these things with me."

5. *Birthorder Position / Vantage of Orientation*

The child finds a vantage of orientation among (or in the absence of) siblings from which to consider the situation of self, others, and the world, and concludes, "This is where I stand and how I must stand amongst others in order to maintain my bearings."

6.   *Genetic Possibility / Self-Assessment*

The child's evaluations consider the physical and mental constituents of the self as among the features of the world. These may include organ inferiorities and anomalies, particular strengths, degree of activity, mental acuity, and any other perceived constitutional capacities for life in the social world, drawing the conclusion that, "These are my personal limits and possibilities for making a place amongst the others."

7.   *Environmental Opportunity / Openings for Advancement*

The child evaluates the situation as including both openings for and limitations to advancement, concluding, "This is what is open to me in life, and these are the obstacles that stand in my way."

Conclusion

In Chapter 1 we set out to orient the reader to the distinctiveness of Individual Psychology's understanding of personality development and dynamics, its field of inquiry, and its method of correction.   In Chapter 2 we presented an overview of the psychoclarity process in lifestyle assessment.

The coming parts of the book address three aspects of procedure: gathering and recording information (Part II: Inquiry); reviewing the gathered material with the client to gain as complete an understanding as possible (Part III: Review and Interpretations); and finally, summarizing the conclusions the therapists and client have reached together (Part IV: Summaries of Understanding).

THE CREATIVE CHILD

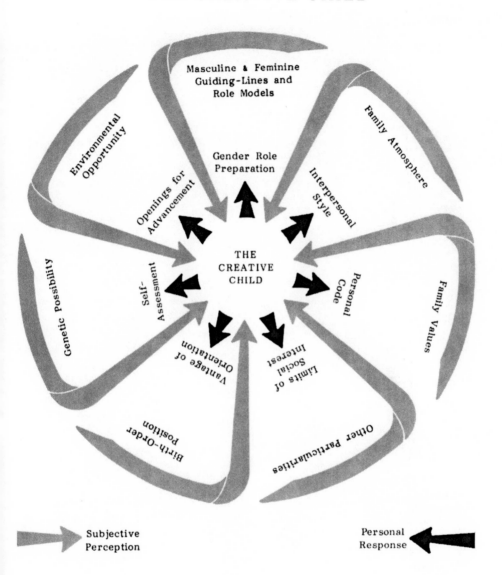

Figure 4.

PART II: INQUIRY

In a considerable portion of neurotics, the turning to a physician, sometimes even to another person, is of itself enough to drive the patient on further. In approximately half of the neurotics, the call on the physician signifies their decision to get better, that is, to give up a symptom which has become superfluous or disturbing. It is this 50 per cent of "cures" that enables all schools of psychiatry to continue to live.

Alfred Adler (p. 336)

Part II is a discussion of items covered in the initial interview and lifestyle assessment inquiries, clarifying the intent of each item and providing examples of the kinds of information elicited in a variety of client responses. (The complete outline of the items discussed in this section appears in the *IPCW*.)

The inquiry is conducted by the recording therapist. After the inquiry is completed, the client, recording therapist, and consulting therapist convene for the review and interpretations and the summaries of understanding. The recording therapist is responsible for the inquiry, both therapists participate in the review and interpretations, and the consulting therapist is responsible for the summaries.

This procedure follows an innovation by Rudolf Dreikurs, (Dreikurs, Mosak, & Shulman, 1952), who used the term "multiple therapy" to designate his model for more than one therapist working with the same client.

Multiple therapy enhances the opportunity for awakening the client's sense of engagement in shaping outcomes. Adler called this sense "social interest," a term he borrowed from William James, mistakenly believing it to be nearest to the meaning of the German, *Gemeinschaftsgefühl*. This German word is now more commonly understood to mean "community feeling" (Griffith & Powers, 2007, p. 11).

Individual Psychology's criterion for measuring the value of any therapy is the extent of the increase in the client's community feeling. Here, in the very structure of the therapeutic encounter, the client is invited to experience community feeling from the start by engaging with two actively interested professionals, and by working with them in a joint effort to understand the private sense of his or her own personal style. In this, the process itself offers a foretaste of the desired outcome.

The client's sense of equal engagement is further enhanced by seeing that each therapist makes individual and distinctive contributions to the work, each from a unique vantage, each in the exercise of a particular set of responsibilities.

In the inquiries, the task of the recording therapist is to enter into a conversation with the client. This conversation is unlike others in that it has a therapeutic goal and is carefully ordered to achieve that goal. The framework provided by the *IPCW*, is therefore intended to guide the discussion, allowing it to unfold in a coherent manner without interfering with its conversational character.

Therapists will use the outlines and items detailed here in line with their own styles and the styles of their clients. For example, if clients move from talking about their children to talking about their siblings, there is no need to interrupt; turn the page and write the sibling information in the appropriate section of the outlines. Or, if the therapist has a sense that it is important to address the question of a client's health before asking for information about the client's parents, this is the sequence to follow. The format we present here, in other words, is to be the *servant* of the conversation, not the *master*.

The initial interview is organized so that the recording of apparently external facts (date of birth, place of employment, etc.) begins a process leading to a consideration of the client's current life situation, presenting problem, and self-assessment in the areas of the three inescapable tasks of life: *affiliation,* in friendship and in the broader community, *work,* and *love* (Adler, pp. 131-133, 297). The client thus moves step-by-step from discussing matters of consequence to all human beings toward addressing issues of *unique personal difficulty.*

The recording therapist is responsible for noting signs of exaggerated agitation, hesitation, lability, jocularity, and so on.

Silences or other retreats from engagement in response to particular subjects must be noted as well. These are important indicators of sensitivity, and they may show themselves at any point during the interview. The recording therapist must also be prepared to offer encouragement; to keep the atmosphere light without making light of the issues; to allow time for the silences; to offer comfort and sympathy as appropriate ("I can see that this is hard for you"); to make self-disclosures when they are called for; and to maintain an interested attitude in the face of anger, or tears by saying (to anger, for example), "I can see how you could feel that way," or (to tears), "Let the tears come . . . . What comes with the tears?"

Chapter 3 covers the inquiry in the initial interview. Chapter 4 addresses the inquiry for the lifestyle assessment. You will note the heading "General Diagnosis" for the first, and "Special Diagnosis" for the second. These are Adler's terms. Our use of them here is intended to draw attention to our following Adler's lead in this area, as described by the Ansbachers (Adler, 1964):

Understanding the specific lifestyle of the patient, his specific problem situation, and the specific significance of his symptoms . . . is accomplished by an initial, tentative understanding of the patient, which is gained through empathy, intuition, and guessing. On this basis Adler formed what we might call a hypothesis of the patient's lifestyle and situation. At one time, when addressing a group of medical men, he termed this step of forming the hypothesis, the *general diagnosis*, and termed the verification of the hypothesis, the *special diagnosis*. This second step consists in a careful interpretation of all the expressions of the patient, to see what bearing these have on the hypothesis. "In the special diagnosis you must learn by testing . . . At first you have to guess at the cause. But then you have to prove it by other signs. If these do not agree, you must be severe enough with yourself to reject your first hypothesis and seek another explanation." (pp. 326-327)

In our experience, these two "steps" (as Adler called them), namely an empathic, intuitive guessing and a "severe enough [willingness] to reject [our] first hypothesis," are the psychological actions essential to following the living movement of another person in lifestyle assessment. These two steps are also the most difficult to teach, probably because they are matters of virtue more than of technique.

The virtues referred to here are not heroic or saintly, but the more ordinary and everyday ones, made possible by our interest in each other, and our imaginative capacity for identification. These include the willingness to extend oneself in empathy and intuition, as well as the courage to guess. Some will lack the former because they are more interested in knowing better; others will lack the latter because they are more interested in being right (and will want more and more data to ensure that they have not overlooked anything before risking a hypothesis). As practitioners gain confidence in what they have to offer, they become more and and more capable of empathy and develop more and more courage to make guesses. Then a further virtue is called for: the humility to acknowledge wrong or useless guesses when they are corrected or rejected by the client.

General Diagnosis

INITIAL INTERVIEW

<u>The Life Situation</u> (Items 1 - 22)

1. Date_____

The date of the initial interview may relate to a particular recent or upcoming event of importance in the client's life, or to an anniversary of a past occurrence of importance. Bear this in mind as you inquire about date of birth, date of marriage/partnership, ages of the client's children and parents, and the dates of significant events in the lives of the parents.

2. Present Age _____

3. Date of Birth _____

The human experience of the present includes an expectation of the future and a remembrance of the past. Lifestyle convictions include the former and rest upon the latter. In any present moment we operate on the basis of our expectations, guided by the wisdom we believe to reside in our memories, our early recollections. The client's present age has an importance relative to these private and (most often) un-understood lifestyle expectations.

- A single, twenty-nine-year-old man said, "Six months from now I'll be thirty. I always thought I'd have everything sorted out by now, and I don't even know what I want." He had set age thirty as a Big Number (see the "Big Numbers," Chapter 6) by which to measure his accomplishments, and felt profoundly uneasy and anxious as his birthday approached.

- A divorced woman said, "I just turned thirty-five. Not only don't I have any children yet, but I don't even have anybody to have children with! By now I thought I'd have kids in school. I can't figure out what went wrong."

- A man, forty, and unhappily married, exclaimed, "I just hope to God the next forty years aren't going to be like the last forty. I've got to make some changes, but I'm afraid it's already too late."

Knowing the period in which the client was born also helps to establish the context in which that person developed a unique style. A sense of what life was like, economically, politically, and socially during the time of a person's childhood can be of fundamental importance to understanding his or her childhood evaluation of the world, and how, and to what extent, the current situation may challenge that evaluation.

- Harold was born in 1946. His boyhood began in the years of recovery after the struggle against fascism and militarism in World War II. He entered his adolescence as the postwar period readjustment, the beginnings of the Cold War, and the ambiguities of the Korean War were yielding to the anomie of the Beat Generation. He experienced the years when togetherness, propriety, and optimism gave way to the Cold War and the Space Race.

- Kristen was born in 1953. Her childhood took place in a time of general well-being unfortunately marred by the cloud of the Vietnam War and affected by the agitation of the civil rights movement. Her adolescence and young adulthood were challenged by the rise of the new Feminism, proliferation of drugs, and the sexual revolution, made possible by access to reliable birth control ("The Pill").

- Eliot was born in 1985. His boyhood was experienced during the Wall Street Boom, the introduction of technology into daily life, and unprecedented affluence. The good feelings of the period were offset by the rise of Islamic militancy, the World Trade Center attacks of September 11, 2001 ("9/11"), and the subsequent Iraq and Afghanistan Wars. The wars continued through Eliot's adolescence and beyond, a time of economic dislocation consequent to globalization and the 2008 crash of Wall Street. In the same period, the Lesbian/Gay/Bisexual/Transgender/Questioning/Queer/and Allies (LGBTQQA) community emerged from the closet into full public view, anti-immigrant hostility increased in some quarters, and political divisiveness was pronounced.

As we move forward with clients (personified here by Harold, Kristin, and Eliot), we consider what their experiences could have been growing up, and how they interpreted the world situation and mood, all of which has a bearing on understanding him or her. None of client's present movement can be clear to us *unless we are thinking contextually.*

4. Name, address, home or cell phone; if female, note maiden name or former name(s); full name of spouse/partner:

Record the client's full name (if female, include maiden name if she took a married name) and the spouse's/partner's full name. If the names are not familiar to you, inquire about their ethnic/racial origin. Ethnicity/race can add to your sense of context. It may provide a framework for thinking about the client's family of origin, values, and religious attitudes. If the client is married, or in an intimate partnership, the names may reveal that this person moved "out" of one ethnic or racial group and "into" another, possibly perceiving the union as a move "up" or "down."

- A thirty-seven-year-old man from a Polish Roman Catholic family had married a Protestant woman of English extraction. He had left his family religion and working class neighborhood and established a social life among the corporate managers with whom he now worked. His family refused to accept his wife. He was troubled by this, but even more troubled by his sense of disadvantage in moving among his wife's wealthy, Protestant friends in a new suburban milieu. He had wanted to enter this world, and was proud of his wife, but he had misgivings, and often felt, as he put it, "in over my head."

Home address may tell something further about a client's sense of social status, sometimes even of social values.

- A woman of thirty-five gave her address in a predominantly Jewish community, adding that she planned to move soon, that the neighbors were "not my kind of people," and that she didn't want her children to attend school there.

- A man of fifty-two gave an address in an area known to be a neighborhood hospitable to same-sex affinities, providing an opening for later exploration of sexual orientation and gender issues.

5.   Does your name carry any meaning for you?

Answers to this question may lead to other subjective self-evaluations of importance.

- A man of sixty-six said, "I have an odd name — one nobody ever heard of. I hate it, but I'm stuck with it. It's Finnish. I was named for the Finnish farmer who owned the neighboring farm. When people learn this, they raise their eyebrows and crack jokes about my paternity."

- A woman of forty said, "My mother named me Jane because she was crazy about the movie star, Jane Fonda. I always resented my name. I didn't *want* to be associated with

Jane Fonda! When I left home, I began to spell my name 'Janne.' Everyone comments on this spelling, which is fun for me. My mother was, and is, enraged by this, but then, she's enraged by everything."

- A man of twenty said, "I'm named for my father and my grandfather and my great-grandfather. This means a lot to me. I know a lot about these men. They were good people. I'm proud to carry the family name, but I dropped the Roman numerals — that's a bit much."

6. Occupation, position, work address, work phone; occupation and position of spouse/partner:

This information further defines the context of the client's objective situation.

- A man of thirty-four was employed as a nuclear engineer. (Begin to consider probabilities: Such a person is likely to place a high value on the understanding and control of details. He may be more cautious and constricted compared to people drawn to other occupations. He may enjoy tasks that involve immediate measures for determining whether solutions are right or wrong. He may appreciate being part of an orderly, predictable environment in which he knows exactly where he stands, and in which there is no doubt about who is responsible for what.) He complained, "I don't have any fun."

- A woman of twenty-nine was an accountant studying for the C.P.A. (Certified Public Accountant) examinations. She is punctual, and carefully groomed. (She may want everything to "add up," and for life to be coherent and predictable. She may have carefully chosen friends in a narrow social circle.) Here is the way she described her problem: "I'm lonely. I never seem to go anywhere or do anything."

- A man of thirty-one was a sales representative. (He may value being his own boss, may be impatient with details, and may enjoy a more free-spending, gregarious way of life than those in other occupations. He may have an appetite for spontaneous camaraderie in an ever-new group of acquaintances.) His statement of the problem was: "I can't hang on to an intimate relationship."

- A man of sixty-two was a restaurateur. (He may be a workaholic whose business is his life. He may find his identity in the spotlight with many patrons who feel special when he gives them personal attention and who flatter him in turn.) He complained: "My wife has gotten interested in a lot of other things, and wants a divorce."

- A woman of thirty-six was a furniture importer. (She may enjoy travel and excitement. She may be artistic, concerned with appearances and glamour. She may value independence, and avoid close ties. She may be self-absorbed both in work and play, and may expect to be the center of things.) Her complaint was: "My boyfriend doesn't treat me right."

Occupations attract people whose lifestyles are congenial to the requirements of those occupations. Choices of occupation can help us to understand the various patterns of style, in terms of their effectiveness and their limitations, the kinds of success they enable, and the kinds of difficulty they entail.

Work address may also yield information concerning style.

- A woman of forty-two worked out of her home as a commercial artist. She valued privacy and self-reliance. She had been unable to form an alliance with a man because she feared she would have to "compromise" her freedom.

- A man of forty-eight was an insurance executive with a fashionable downtown office. He enjoyed the fast life of the city, had flirtations and affairs, went drinking with his men friends after work, and complained: "My life is empty."

Noting the spouse's or partner's occupation may also provide clues relating to the current difficulty.

- A man of forty-five stated that his wife was a buyer in a department store. He complained, "Ever since she went back to work six months ago our marriage has fallen apart. She's never home, and nothing gets done. She has a new circle of friends, and I'm not included. I don't know what to do."

- A woman of thirty-two was married to an attorney who said his practice wasn't enough for him. He wanted to go into politics, a life she hated: "If they aren't already corrupt, they get corrupted. Ever since he got bitten by the bug, he's been a different person."

If a client moves immediately to discuss the details of the presenting problem before you have had an opportunity to gather the basic material in The Life Situation section of the lifestyle inquiry, you can say, "It appears that this part of your life is the most troublesome. We're going to explore it together in detail, and I think we'll be able to make more sense of it if we first examine the context in which it's happening. Let's take a few moments more for the basics so that I can get to know you better; then we'll move to that subject. Is that all right with you?" We have never had anyone say no; people appreciate having the opportunity to proceed in a systematic way. The structure may even come as a relief to a client's chaotic way of thinking and talking about troublesome issues.

When there are other digressions from the topic at hand, one must use tact and a friendly confidence to guide the conversation back to the subject. "I can see that this is important to you. I'd like to come back to it when we have more time to explore it." Such an approach usually enables the client to return to the subject.

7. Preferred phone and mailing address (home or work):

A client may say, "Don't call me at the office." This person may see the office as gossipy or political, an environment in which people must be cautious not to disclose their personal lives. (This is often the issue when people whose visits could be covered by group insurance do not want to file a claim for psychological services.)

Someone may say, "Don't call me at home." It may be that this client's spouse/partner is unaware of your being consulted; the statement may imply that a degree of trust is lacking in the marriage/partnership. It is advisable to inquire about this, to know the reason for the secrecy.

If a client is still living at home with parents (or a parent), this may be relevant to the presenting difficulty. It may indicate a hesitation of the client's with respect to being "on my own," completing school, or identifying and pursuing an occupation.

8. Marital/partnership status (age of spouse/partner, how long together, anniversary; previous unions of each; or, current love relationship, how long together, level of relationship):

In the review of the life tasks the issues of love and intimacy will be addressed in some detail. Here we are concerned with the objective situation. If there was also a previous union (or unions), say, "Tell me in a few words what you think went wrong with your previous relationship(s)."

> - A man of twenty-seven responded, "I know my former wife would tell you something different, but I don't think she ever grew up. Whenever anything went wrong she got her mother tied up in it, and I'd always come out the bad guy. She never tried to solve anything with me. She'd just withdraw or call 'Mommie.' I finally called it quits."

In this connection it is important to ask who initiated the divorce or separation, and what prompted him or her to act.

Ethnic/racial background, religion, social class, or other contextual features may be at issue. If there is a marked disparity in the ages of the partners, this may be a matter to consider. Of course, any disparity can be a source of enrichment for a partnership or marriage, or, depending on the meanings the partners assign to it, it may also become a source of trouble.

- A man of fifty-one was in his second marriage. His wife was twenty-nine. He had adult children from his first marriage. Now his new wife wanted a baby. During their courtship he had said that he would want a child with her, but secretly hoped that she would give up the idea after they were married and caught up in their new life. Contrary to this hope, she was pressing him and threatening to leave. He wondered, "What's wrong with me? I like children, and I love her. Why am I unwilling to go through with this?"

9. Children (names, ages, sex; disabled and/or deceased children; difficulties of conception; other pregnancies, adoptions, fostering; children from other relationships):

The importance of this information may not show itself immediately, but the data fills out a picture of the client's life experience. Later you may find a significance in the dates of the children's births, or the timing of the pregnancies, or the dates of other events connected with the children. (A correlation may show itself at Item 24, p. 71: "When did the problem start? What else was going on in your life at that time?")

- A man, forty-nine, was having marital difficulties and was considering divorce. The trouble had begun fifteen years before. We asked, "Wasn't that about the time Jennie was born?" It turned out that he had felt overburdened by the birth of this child, their fifth, born five years after the next oldest. He was then having financial problems, hadn't wanted more children, and felt that his wife had "tricked" him into the pregnancy. From that time on he had felt resentful and betrayed, and as if he were all alone in the task of making a living and providing a life for everyone: "Now I'm almost fifty, and it's time for me to have my own life, before it's too late."

- A woman of thirty-eight was the mother of three children. Six years before, while she was away from the house, her husband, a physician, was looking after the children. He had stepped into the house to get a drink when the middle child, then four, ran into the street and was hit by a car. Even though the father then rushed to the scene of the accident and saved the child's life by dint of his medical training and competence, the client felt unable to trust him ever again, and had come to see him as vain and uncaring.

You may learn that the client married after the beginning of a pregnancy. Many firstborn children "arranged" the marriages of their parents unawares. Sometimes the result is a good arrangement, sometimes not.

- A man of forty-two had "had" to marry his wife. He had also had to drop out of school and give up his ambition to become a lawyer. He was angry and resentful: "I did that for her, not for me. She's never appreciated it. The least she could have done for the past twenty-two years is show a little gratitude. I saved her reputation, and it cost me my life. I never would have married her if it hadn't been for the pregnancy."

The birth of a child may have other consequences.

- A stockbroker, thirty-six, had been the baby of his family and his mother's favorite. He married a much younger woman who had been doted on by a grandfather, and who expected her spouse to pamper her. The relationship was stormy, with neither of them feeling adequately served by the other, but there was a lot of money for each of them to spend, and they managed to stay together for ten years. Then she decided that she didn't want "to miss out" on the experience of having a baby; when the baby arrived, she was absorbed in catering to the infant. The client felt abandoned and ignored, and said he "couldn't handle it."

Adoptions, fostering, miscarriages, stillbirths, abortions, children who are developmentally disabled or placed in institutions, and the death of an offspring should be noted. So also should technical methods for achieving conception, or any other sexual difficulty, which may be an indication of diminished compatibility, trust, and confidence in the union.

- A couple, unable to have children after several years of trying to conceive, adopted two children in rapid succession. Within five years they were considering divorce because, as the woman put it, "We argue all the time, we don't enjoy each other's company, and we don't have anything in common."

An exaggerated idea of the responsibilities of parenthood can also be at the root of a difficulty in conception.

- A thirty-year-old man was in despair over a marriage in which "everything is good, except the main thing. We can't have children."

He and his wife had consulted fertility experts, and she had been told that she had an "infantile uterus." They had agreed to adopt, but when the time came for a decision she had faltered and offered a host of reasons for delay. He felt betrayed, and was considering divorce. The whole line of his ambition as a man was in the direction of being a successful father, a task at which (to his mind) his own father had been defeated by his mother's depressive personality.

At our urging, he brought his wife with him for a meeting. She was reluctant, because as she put it, a consultation with a psychologist was a sign of "weakness," and she hoped her mother would not learn of her having resorted to it. We saw an opportunity, and asked them not to discuss children in any connection for six months while we worked with them to find out how two people, so clearly fond of each other, were experiencing so much difficulty. They agreed.

Within six months she was pregnant. She had come to see that the feminine guiding line (see Chapter 6) set by her mother had left her with the conviction that motherhood was an heroic task, and that only when she was equal to taking care of anything and everything on her own, without anyone's help, would she be equal to it. When she was liberated from this constricting idea, she was able to conceive. They are now the happy parents of three healthy children (and he has greater respect for his father).

Also, in the inquiry about children it is important to note any special alliances between either of the parents with a particular child, perhaps especially a troubled child.

- A woman of forty-eight complained that her husband was so involved in the life of one of their grown daughters that she felt overlooked and disregarded. He in turn accused her of "selfishness and callousness" regarding the daughter's continuing difficulties, and husband and wife were estranged.

10. Level of education; military service; religious affiliation:

This item elicits further information regarding a person's history and range of experience. A client may offer important evaluations as well as provide unexpected information while reporting the facts.

- A man of thirty-three said he had an M.B.A. (Master of Business Administration), and that he was the first person of his family to complete any degree, saying, "I've felt a little uneasy around my father ever since I finished college, and now that I have a master's degree it's even harder." He told us that he felt as if he had "surpassed" his father and his family's working class position: "I feel like a misfit in my own family."

- A woman of thirty-eight, with an M.S.W. (Master of Social Work) degree, said, "Of course, I only got that degree after I left the convent."

- A woman of forty-one had dropped out of school and made a career in advertising. She had been promoted to the senior management group in her company and felt unequal to the position: "I'm not only a woman; I'm an uneducated woman."

Military service (or lack of it) may be reported matter-of-factly, or your asking about it may stimulate further revelations.

- A thirty-six-year-old man had enlisted at age 18 to get away from an intolerable family situation: a drunken Father who brutalized Mother, who wouldn't do anything about it.

- A thirty-seven-year-old client had another reason for enlistment: The primary family value and training was a fundamentalist pacifism against which he volunteered in an act of defiance.

- A man of thirty-four said he was a *"conscientious conceptor,"* explaining that he had married and had a child immediately to avoid the Vietnam draft. He felt he had "gotten away with" something, and suffered with an uneasiness that he had not gone with his friends, some of whom had been wounded or killed. "What kind of man am I?" he wondered.

In addressing religious affiliation we pose the matter this way:
"Tell me about your religious orientation and practice, if you have
one."

- A woman of twenty-seven said, "My family is Jewish. It
was a social thing for them, and I hated their attitude. I
decided to do it right if I was going to do it at all, and I keep
a kosher house now. In fact I chose my husband, in part,
because he was so strict about it."

- A man of forty-one said, "I was brought up Catholic, went
to Catholic school, the whole bag. I got out as fast as I could.
I leave it alone, and it leaves me alone. I'd like to be a
Presbyterian, but I haven't done anything about it."

We asked what made Presbyterianism attractive to him, and he
said, "It isn't emotional. They're not all hung up on guilt."

- A woman of twenty-nine said, "I'm Catholic. Church, and
everything related to it, was important to all of us when we
were growing up. Going to church was special. We were
together, we all looked our best, and there was the music,
and the flowers. It was the only beautiful thing in my life."

- A woman of twenty-eight reported, "My father was a
fundamentalist minister. We couldn't go to movies, watch
TV, play cards, or smoke. It's really funny in some ways.
We teenagers figured out that the only thing we could do was
have sex: They never even talked about that, so we must
have decided that it wasn't forbidden! We were not part of
the town. They looked down on us; we looked down on
them. I got out of there as fast as I could. But, even though I
don't belong to a church, whenever I feel down I go to the
nearest church to pray. I don't know why I do that."

11. Parents (original marriage/partnership; if parents separated or
    divorced, client's age at time of disruption; same-sex parents;

note parents' ages at client's birth; previous marriages or partnerships of each parent; other adults present in the pre-adolescent household; if parent(s) alive, note age, state of health, current situation; if deceased, note age, date and cause of death; client's age at parent(s)' death(s):

With this subject we begin to form an impression of the masculine and feminine guiding lines defined and explored in Chapter 6.

- A woman of forty-three was executive secretary and office manager for the owner of a new company. She had gone to work for him five years before, on a temporary basis. This was when her husband, then forty, was recovering from the first of a number of stays in hospital for a degenerative disease that limited his ability to work. Since then the company had grown; she was regarded as "indispensable" by her boss, who paid her well, relied upon her, and took a personal interest in her life and her problems.

She reported that her father died when he was sixty-five, eight years before. "He was healthy all his life, and went to work no matter what. I doubt if he missed a day. He had a stroke just one month after he retired. My mother is still alive at sixty-eight. She developed a severe digestive problem after Dad died — no diagnosis. It worries me. She came to live with me about six months after Dad's death; I didn't want her living by herself. You know, she was one of those women who never worked. My father wouldn't hear of it.

Only once, when he was on strike, he let her work nights in a restaurant, but he hated it, and so did I.

This woman came to us complaining of intermittent depression, agitation, and sleeplessness. It was not hard to see that her expectations for a woman's place in the world were not being met: A woman could expect to be supported and cared for by a man; a woman was not supposed to work at all (much less be "indispensible"). She laughed when we pointed out that anyone growing up as she did would be likely to feel disoriented, perhaps misled, and probably resentful. From that moment on she began to feel better about herself, more accepting of her unforeseen responsibilities, and more free to enjoy the self-esteem that she had gained with her employer's encouragement.

> - A man of twenty-nine reported, "My parents divorced when I was eleven. I never saw it coming. My stepmother, Joyce, married my Dad when I was twelve. My stepfather, Harry, married my mother when I was sixteen. I think Mom waited to see what Dad was going to do, and whether it was going to work out with him and Joyce. She really wanted him back. I think he already had something going with Joyce while he and Mom were married. I always thought so, and I always resented Dad and Joyce. They really wrecked my life. I still have a hard time being civil to my Dad."

This man had come to us because he had broken his engagement three times from the same woman, who remained devoted to him. He said that she was "perfect" for him, but that "I lose my nerve at the thought of the responsibility of marriage."

12. Were you told you looked like anyone? Was this important to you or others? In what ways?

> - A woman of fifty-six said, "From the time I was a little girl people said I look exactly like my father. I was a girl! I didn't want to look like my father who had a big face and a big red nose and bushy eyebrows. It was disturbing to be told how much I looked like him."

> - A woman of thirty-three said, "I looked like my mother. I was flattered to be told this. My mother was a beauty. Of course, I'm *not*, really, but since I'm the only granddaughter on both sides, you might say people courted me. They complimented me and fussed over me. I enjoyed it!"

- A man of sixty-four said, "Apparently I didn't look like anyone in the family. I was sort of an ugly duckling. When I was a kid, I always wished I looked like my dad, because in a way, I felt like kind of an outsider in the family."

13. Siblings (list client and siblings by first names from oldest to youngest, placing the client, deceased siblings, and Mother's other known pregnancies in their ordinal positions; using the client's age as a baseline, note the difference in years between the client and each sibling; in parallel columns list step-siblings, age differences between step-siblings and the client, and the client's age when they entered the family; note the present whereabouts and situations of each sibling):

Here we establish the client's position in the birthorder of the children. It is important to know not only about surviving siblings, but also about any miscarriages, deaths, stillbirths, or siblings placed in institutions. Following Adler, we often discover that a healthy child, born after such events and difficulties involving earlier-born siblings, comes into the world as a revelation, and may have grown up with an aura of invincibility.

Birthorder is not to be mistaken for a determinant of the individual's style of living (see Chapter 8); birthorder information is important in that it gives us an idea of the particular vantage from which the child viewed and assessed self, others, and the world. As Adler put it:

There has been some misunderstanding of my custom of classification according to position in the family. It is not, of course, the child's number in the order of successive births which influences his character, but the *situation* into which he is born and the way in which he interprets it. (p. 377)

For this reason empirical studies, inspired by Adler's emphasis upon the significance of birthorder, have often been inconclusive and unsatisfactory. In no other area of this inquiry is Adler's general maxim more to be taken as an imperative: "*Alles kann auch anders sein*" (i.e., everything can also be different) (p. 194).

- A woman of twenty-three came to see us because she didn't "know what to do" with her life. She was one of two children, with an older brother. As she reported it, "I was the baby. My brother Jim is three years older; he's a lawyer."

- A woman of thirty-six complained that she couldn't get along with men. She had been the only child of her parents until her two brothers were born, one when she was fifteen, and the other when she was seventeen. She reported that she was embarrassed when she was in high school and her mother became pregnant. "How could she have done that to me?" She also reported that her mother had always had "female trouble," and that it was after an operation to rectify this that Mother had been able to conceive again. This woman did not enjoy sex and had the idea it was "the only thing men are interested in."

- A man of forty-two was in torment over his marriage to a woman seven years younger, who had told him she didn't love him, and who was having an affair. "She's so naive. I'm afraid to think of what would become of her if I left." He was the firstborn of three, with a brother three years younger and a sister six years younger. He was his mother's favorite. "Dad worked hard and wasn't around much, so it was Mom and I, and the little kids, and I helped her out. I went to work when I was twelve. My hypertension was diagnosed when I was thirteen years old."

- A woman of thirty-one was the first of four. "The only one who bothered me was my brother Tim; he was eleven months younger than I. He was always getting into trouble, and fighting, especially with me. If he hadn't been so bad we would have had a nice family." This client consulted us because she was having problems with her son.

14. When you were a child, if anything could have been different, what would you have wanted it to be?

This question may uncover something of the direction of a person's hopes, striving, and ambitions. It may also elicit information about disappointments in life, and, perhaps, the client's claim to being a victim of circumstances, with a right to bear a grudge against others and the world. For Adler (1964), goal-directed movement can only be understood "as the individual's efforts to secure for himself what he interprets, or misinterprets, as success, or as his way of overcoming a minus-situation in order to attain a plus-situation" (p. 181).

- A woman of fifty-five was an only child. "I would have liked to have had an older brother. It was so hard to always have to be the ideal daughter, to have to be perfect. My parents doted on me, and I never had a moment's peace. With an older brother, I would have had someone to share all that pressure. Maybe I could have done something on my own. Maybe I could have had some fun."

- A man of forty-one was the firstborn and only son, with six younger sisters: "I would have wanted more privacy and more space. I never had a room of my own; there was always a baby in it. My parents went on having children even when they couldn't take care of the ones they already had." He felt deprived then and continues to feel deprived. He has struggled all his life to create his own world. He owns his own business, and has a large, comfortable house, which he maintains as a bachelor. He came to us because he found it "impossible to keep employees. I can't work with anyone."

- A man of thirty-eight reported, "I would have wanted my father to be able to hold a good job. He was always out of work, and there was never enough money. The fights Mother and Dad had over money never ended. They had it tough, and they made it tough on each other. I swore that would never happen to me." He was a successful politician. He always knew someone with whom he could make a deal, get a job, get ahead. He was rarely home, and his wife wanted a divorce. He was amazed: "Why would she want a divorce? She's got everything she could possibly want! What more does she want? I'll give it to her!"

15. Are you currently under a physician's care? What for? (Note type of problem and any signs; if the client's presenting problem is a physical symptom, see Item 23, p. 69):

A psychological investigation is not complete without a thorough inquiry into any physical problem. Whether or not it relates to the presenting complaint, it has a meaning to the client. You should also be prepared to inquire about any signs of a possible physical disorder, whether or not the client refers to them. A client's failure to refer to an obvious disorder is significant in itself, and should be understood.

- A man of thirty-seven reported that he was a compulsive overeater. This was connected to his presenting complaint (to which we turned at this point in the inquiry). His physician had sent him to Overeaters Anonymous. He said, "When I work the program, I can lose weight. A month ago I dropped out of the program, and I'm gaining again. It goes in cycles, and seems to depend on two things: How busy I am at work, and whether or not my wife shows any interest in me." He appeared to be moderately overweight, looked tired, and showed little energy.

The physical effects of an illness must be interrupted before they crystallize into permanent damage. If a client reports a physical symptom and has not consulted a physician, we insist upon a medical examination. Further, any symptom, indeed any illness, *may serve the purpose of psychological expression*, which we will want to help this person understand. (See p. 71, below, regarding Adler's concept of "organ jargon.")

16. Are you taking medication(s)? (Purpose, name, dosage, how long; effects and side effects):

Note the purpose for which any medicine has been prescribed, the name of the drug, the dosage, how long the client has been taking it, and its effects and side effects. Sometimes side effects come first to the attention of the psychotherapist. Further, it is not safe to assume that the prescribing physician will always be alert to their occurrence, or even aware of the frequency with which such effects may be associated with the drug. A consultation between you and the physician may be called for, and you will have to obtain the client's written permission for this purpose before you proceed.

> - A woman of twenty-four, barely more than five feet tall, had consulted a physician six months before our meeting in an agitated and depressed state over the deterioration of her marriage. The physician prescribed an antidepressant drug with a tranquilizer. Since then she had gained over twenty-five pounds, and was unhappier than ever. We turned to *The Physicians' Desk Reference* and found that weight gain is a frequent side effect of the drug she was taking. Just knowing this was an immediate source of encouragement to her. She consulted her physician, and upon stopping her medication, her weight gradually returned to normal, while psychological counseling restored her self-confidence.

17. Date of last physical examination:

We routinely recommend a physical examination if the client has not had one within the past year. Even when there was an examination during the past year, we may recommend another if the person has since experienced catastrophic stresses. If the client does not have a personal physician we refer to a local clinic which does a thorough exam on a sliding fee scale. The clinic sends a complete report directly to the client, who may then review the results with our assistance.

- A woman of forty-six said, "I don't want to go. If I do, I just know they'll find something wrong with me." Her mother and sister had died of cancer, and she suffered with the expectation that she would also have the disease. We were able to persuade her to consider the possible damage her pessimism and fatalism could do to her general health. A physical examination revealed a mild nutritional anemia that was treated successfully with iron and vitamin supplements. She began to see how her discouragement had kept her from maintaining a proper diet, leaving her more vulnerable to any illness, including the illness she dreaded.

18. Name, address, and specialty of physician(s) being consulted:

Be prepared to give a referral to an appropriate specialist, if this appears necessary.

- A woman of thirty-eight was fearful of many things, including dentistry. "My teeth and gums are rotting, and I'm terrified of going to a dentist." We referred her to a dentist who is psychologically sophisticated, and who was able to establish a relationship with her one step at a time, beginning with simple instructions on dry brushing and the gentle use of unwaxed dental floss, and going on to the painless process of correcting her bite by equilibration. He went with her to the offices of another dentist, who was an expert in root canal surgery, and stayed with her through the procedure. She became fearless, and enthusiastic about the restoration of her mouth, even to the point of wearing braces!

19. General health (self-identified sexual orientation; level of sexual activity; appetite, sleep, energy level, exercise; use of alcohol and drugs, including recreational drugs; use of caffeine, tobacco; note any diagnostic signs):

We ask, "How would you describe your general state of health at this time in your life?" This gives us the opportunity to inquire about specific health issues, if the client has not already addressed them in responses to the preceding items.

> - A man of forty-two made an appointment to discuss his career plans. He was a corporate manager, considering an entrepreneurial venture in partnership with a friend. He made no reference to marital difficulties until we asked about his level of sexual activity and degree of satisfaction in that part of his life. He responded, "My wife and I haven't had sex in six years, at least!" It later unfolded that his idea about occupational change was connected to his hope of being able to earn enough to satisfy his free-spending, pampered wife, a "hope against hope" that, if he provided better, he might regain her respect, and with it her love.

Sometimes a client's worries about his or her own patterns of drug or alcohol use will come forward here. Inquire whether any others involved in the client's life worry about it as well. Active substance abuse of any kind interferes with the possibility of psychological investigation. Be alert for every subtle hint and any small cues in this area, since the phenomenon of denial is inherent in a pattern of abuse. Often you will be unaware of any trouble until it is blatant.

> - A man of twenty-seven, whom we had counseled several years earlier, and whom we had seen intermittently since then in connection with his career difficulties, called the office on a Friday evening to describe himself as suffering with a severe panic attack. We engaged him in conversation, and detected a very slight slurring of his speech. We asked how much he had had to drink in his efforts to calm himself down, and he admitted to having consumed a six-pack of beer. We pointed out that alcohol was likely to increase rather than diminish extreme anxiety feelings, and asked whether he would be willing to stop drinking alcohol for the remainder of the weekend, to use warm milk to induce drowsiness, and to call us on Monday to report on how he was doing.

On Monday he called to say that he felt much better, that he had talked with various members of his family during the weekend, and that he had decided to enter a hospital with an inpatient alcoholism treatment program. He did this, and subsequently has made good use of the Alcoholics Anonymous program, and better use of further psychotherapy.

In the discussion of general health, as in other matters covered in the inquiry, it is important to note if the client is aware of any marked change in condition, either positive or negative. If so, we pinpoint, as closely as possible, the period when it became apparent, then ask, "What else was going on in your life at that time?" By asking this question we aim to identify what Adler (p. 296) called the "exogenous factor," that is, an event in the client's life which facilitated, provoked, or invited the change.

The significance of the exogenous factor in understanding the client's difficulties cannot be overestimated. It may lie in an event that, to all outward appearances, would seem relatively unimportant, such as a birthday, a promotion, or a move. The meaning of the exogenous factor will only be understood from within the context of the client's private convictions, expectations, and values. It can tell us a great deal about the limitations of the lifestyle, for it typically relates to a challenge which the person felt and feels somehow unprepared to meet.

- A man of thirty-four had muscle spasms and aches in his right arm. They began about four months before he came to see us, at the time of his marriage. He felt burdened by the responsibility of having a wife, who (in his picture of what was required of him as a man) would expect him to be her "strong right arm." [This is an example of "organ jargon," or what the bodily symptom is expressing on the individual's behalf that the client is unable or unwilling to express verbally; see pp. 62 and 71 of the present text.]

20. Previous counseling/therapy (when, reason, type, with whom, how long, outcome):

The purpose of this item is to identify previous times of crisis in the person's life, to discern whether there is a pattern to these crises, to assess "therapy readiness" (the person's expectations of the usefulness of therapy), and to learn what this client has found to be either encouraging or discouraging in a therapeutic relationship.

Here note also whether or not the client has come to see you while still in a therapeutic relationship with someone else. In such a situation there are several possibilities that must be addressed. The person may have come at the recommendation of the other therapist, or may have suspended work with the other therapist for the duration of the work with you; this would, of course, be acceptable. For a client to continue working with another therapist while consulting you, however, is not acceptable, and you must make it clear that a decision is required before you can proceed. It is not to a client's advantage to work with two therapists who are not working as a team in multiple therapy. Differences in approach and emphasis become sources of confusion at best, and at worst defeat the collaborative nature of therapeutic work, with the observations of one therapist used to negate the observations of the other.

> - A man of forty-four had been in therapy off and on for nine years before he came to see us. He had worked with a Jungian therapist, and a Gestalt therapist; had studied Neurolinguistics extensively; had gone to a Freudian analyst for several years; and was currently working with an eclectic therapist. He felt he was not getting anywhere, and told the therapist that he had decided to consult us in an initial interview and that if he got something out of it he planned to terminate his work with the other therapist, who agreed to this plan; as did we. About the outcome of all his previous therapy he said, "I've learned a lot, and gained a great deal of insight, but I still feel that there's something fundamentally

wrong with me that's never been resolved." We said, "Maybe that's what's wrong." [See "The Case of Harry," Chapter 5, pp. 176-177.] He was startled, and asked what we meant. "Maybe your idea that something is fundamentally wrong with you is your mistake, even a form of vanity, a special idea of yourself as handicapped, having to struggle more than others to achieve as much as you do." He agreed to a lifestyle assessment, which revealed the foundation for this idea: the hostility of his mother and his maternal grandmother (with whom the family lived) toward his father, who came from an ethnic group which the two women regarded as inferior, who therefore regarded him, his father's son, as also inherently inferior.

21. Referral source:

If we do not already know, we ask, "Who referred you to us?" (We typically learn this during our first telephone contact.) Knowing the referral source provides information about the person's network. If clients are unwilling to confide this information to us we do not make an issue of it. We find that as trust develops they will come back to the subject at a later time, sometimes even later in the initial interview.

   - A woman of twenty-four did not tell us how she came into contact with our practice until we had seen her several times. Then she said, "I didn't tell you who referred me because I didn't want you to know I was dating Jim (one of our former clients). I thought you would disapprove." Jim was twelve years younger than she and divorced.

22. Emergency contact (name, address, phone, relationship):

At the beginning of our work together we ask the client to give us the name of someone with whom we have permission to make contact in case of an emergency. This can be important in the event of a suicide threat, or a need to arrange a stay in hospital.

- A woman of twenty-seven called us in the midst of a paranoid terror. We agreed with her that she should be admitted to a hospital, and called our psychiatric consultant to make the arrangements. We then told her that we would ask her contact to be with her until the admission could be scheduled. The contact was able to spend the afternoon with her and take her to the hospital.

The Presenting Problem (Items 23 - 30)

23. What brought you to see us? If a physical symptom, ask the client to describe its location, size, shape, sensations, intensity, and frequency; such descriptions may reveal the purpose of the symptom (the "organ jargon"):

The question, "What brought you to see us?" may seem superfluous at this point, coming as it does so far into the interview. However, it is often a relief to a person to answer other questions first, and the relationship established in covering the earlier questions makes this one all the easier to address.

- A man of thirty-seven, close to his thirty-eighth birthday, said, "Few things in my life are constant. I am one of the most talented people I know, but my talents are stretched in so many directions that I feel I am spread thin. I seem to spend my whole life maintaining all my talents instead of focusing on one and moving it forward. I also do this with women.

"The second thing is that I am afraid to express myself. Can you imagine a professor who is afraid to express himself? Isn't that nuts? What I mean is that I'm afraid to take a stand on anything. I hear myself saying, 'Now let's take a look at the other point of view here,' when I know damned well that the other point of view is horseshit that can't be defended intellectually.

"The last thing is that once I've accomplished something in an area, I think to myself, 'O.K., now I've done that. Why should I do more of it?' And I find myself moving on."

It turned out that our client was fighting against a rigid, limited, and inflexible masculine guiding line (see Chapter 6), which threatened him with the idea of intellectual sterility. When this man was born his father was thirty, an autocrat and a self-made businessman full of braggadocio who ruled his household by browbeating his wife and children, and by insisting upon having things the way he wanted them. He claimed this by right, as the provider, and Mother went along, as if in awe of him. Then, when the client was eight years old, Father's business failed as the apparent result of his refusal to accept advice about a shift in the economic climate. Father was never the same; Mother never understood or accepted the change in the family's financial circumstances, and complained about it constantly. The client saw the tables turned between his parents, and ascribed it all to Father's obstinate determination to stick to things and prove himself right.

Some people are focused and articulate in response to this question, and spell out their difficulties in an orderly way; others present a torrent of undigested ideas, complaints, and experiences. In a case of the second kind, it is up to us to guide the client toward the setting of priorities, and toward an organization of the issues. Occasionally people do not know how to answer, but even these tell us something in response to this query, from reports of a general uneasiness and anxiety about life to accounts of things that are "no big deal," but somehow remain troublesome. Others tell us that they came in under duress from a spouse, a boss, or a parent. We want to know what provoked these concerned persons to apply this pressure.

The presenting problem may refer to the chronic and difficult behavior pattern of a person who has long been aware of it, and is at last prepared to sort it out. On the other hand, it may be a situational crisis for a person who is asking for short-term help to manage it.

Some of our clients get what they want from the initial interview, and that is the end of their therapy. Others choose to complete a lifestyle assessment, at which point they may decide their work is done. Still others continue their work with us in order to reconsider the implications of the mistaken ideas isolated in the lifestyle assessment.

24. When did the problem start? What else was going on in your life at that time?

Here we have another opportunity to isolate the exogenous factor and, perhaps, to make use of Adler's concept of organ dialect/organ jargon (Adler, pp. 222-225; for a further discussion of organ jargon, theory and practice, see Griffith, 1984). In the cases below, severe back problems serve to express clients' psychic pain.

- A woman of thirty-eight suffered with chronic back pain. She reported that the trouble began when she was a senior in college. She had wanted to head for New York and become an actress, but her parents insisted that she go on to nursing school, arguing, "You'll have something to fall back on. You can always get a nursing job."

- A man of forty-eight was experiencing back spasms. He had lost his job in a corporate takeover two weeks before he came in. "Until now, I've felt that this is a big world, and there will always be a place for me. Now I'm on the street with no place to go, and no fallback position."

25. Have you noticed a pattern?

Sometimes a pattern has been established earlier in the interview, and the inquiry at this point is unnecessary. If not, the question may represent a new idea to the client.

- A woman of twenty-nine said, "It's the story of my life. No matter what I do, my father objects, criticizes, and puts me down endlessly. I have an idea it's to get at my mother and doesn't have much to do with me, but that doesn't help. It also doesn't help to have Mother support me, because she won't do it in front of my father. She just pretends to go along, and lets him do this to me."

26. What happens as a consequence of having this problem? What does it enable you to do? What does it prevent you from doing? Who else is affected by it? Who suffers as a result ("the target")?

Take these questions one at a time. It is always best to maintain the conversation without allowing it to become an interrogation. You may get the answer to the second question in this item from the client's response to the first one, and not have to go further. What we are after here is the function of the symptom, and its possible *target*.

- A man of thirty-four said, "When I have headaches my wife is very solicitous. She has taken this very well. I usually have them when I get home from work, and she makes me lie down and take it easy."

- A man of forty-seven responded, "When I blow up at work, it's very effective. I have a reputation for being tough, and I get a lot of cooperation when I begin to get impatient with my people. But I know that it's no good; I also get a lot of complaints about it."

27. What have you done about it until now?

Here again, we may already have an answer to this question. If not, we want to learn what the client has done, or tried to do, about the symptom, complaint, or presenting problem. This may help us to find the *private value* of the difficulty, and identify the *hidden meaning* (Adler, pp. 92-93) the person has for maintaining it.

- A woman of twenty-seven, still in competition with her

sister, said, "I've tried to point out to her how wrong she is, but she won't change. She seems determined to cause trouble for me with Mom and Dad."

28. How do you explain this situation to yourself?

The person's own assessment of the difficulty may reveal the question of self-esteem that is involved.

- A man of fifty-three presented the problem of not getting his work done on time. He said, "I have too many people to manage. I get there early every day and stay late. When I get home in the evenings, I work. And I'm not only doing my own job. I pick up the pieces for everyone in my department. No one does the work as well as I do — they all say that. But then there's not enough time to do what I have to do."

- A woman of thirty-four presented the problem of procrastination. "I can't decide what to do first, or how to do it, so I put things off . . . and off . . . and off. Then I have to scramble to get the job done before the deadline. It feels absolutely great to go in with my work on time! But I'm in agony thinking about it for days, weeks, sometimes months before I do it."

29. How would your life be different if you did not have this problem? What would you do if you were completely well (Adler, p. 332)?

Dreikurs referred to Adler's question as "The Question." He incorporated it as an integral part of any initial interview. As Dreikurs (1967) explains:

The examination begins with an inventory of the subjective symptoms and complaints of the patient. [In the format of this text, Items 23-30.] They represent his *subjective condition*. The next step is an investigation of the *objective life situation*. This consists of a careful

examination of his participation at work, in social
contacts, and in his affectional relationships, uncovering
existing conflicts and dissatisfactions. [In the format of
this text, Items 31-36.] The connections between the
subjective condition and the objective situation is
established by the question, "What would you do if you
were well?" The answer will reveal the dynamics of the
existing psychopathology. (pp. 105-106)

The Question is also of great importance in differential diagnosis
of physical symptoms. Dreikurs (1973) specifically pointed to this
in a discussion of the psychological purpose of physical symptoms:

[The technique] consists of the "question." The patient is
asked what would be different in his life if he were well.
The answer indicates against whom or against what
condition or situation the symptom is directed. . . . [The
answers] indicate why the patient is sick, if the illness is
entirely neurotic, or that he uses a physical ailment for
such purpose. Such neurotic superstructure may be
revealed in the case of an actual organic pathology. (pp.
114-115)

The following three examples will help to illustrate the
usefulness of this technique. The first represents neurotic use of an
organ weakness in the circulatory system, without any diagnosis of
organic disease or damage. This is a single man of thirty-one, who
has been medically treated for migraine:

- If I didn't have these headaches I'd go out more socially.
I would have more time for myself, and I'd do things I enjoy.
Now I use my spare time to recover. Weekends are the
worst. Some Saturdays I just lie in bed. If I were able to go
out more socially I'd be able to find a woman.

The second example is of a neurotic superstructure over a
diagnosed organic disease process. This is a young married woman

with recurrent episodes of ulcerative colitis:

- My house would be organized and decorated. We could entertain. My husband says that he's embarrassed to bring people home as things are now. It's funny. My bedroom is finished, but after five years in the house, the living room and the den aren't. I used to cook and throw parties, but not any more. Now I just "stew." Oh, I get things done, but only by the skin of my teeth, and never under pressure. That just puts me to bed!

Finally, an example of suspected neurotic behavior ruled out by the diagnosis of a physical ailment. This is a young couple presenting a marital disturbance brought on by the husband's suspicion that his wife has lost interest in him. When he sits with her after dinner in the early evening, and tries to have a conversation about events of the day, she falls asleep. Aside from this, they both describe the marriage in positive and affectionate terms. Here is part of what she said:

- I don't understand it, and I'm embarrassed to talk about it, because I can see how he must feel. I'd feel the same way, if the shoe were on the other foot! I fight to stay awake, and I want to talk with him, but as soon as things get quiet and relaxed I'm gone. It's not only with him. I fell asleep on the basement stairs once, and woke up on the floor at the bottom. It's a wonder I didn't break something. And it's misery for me to ride the bus! As hard as I try not to, I'll just go straight to sleep as soon as I'm comfortable, and as often as not I miss my stop.

Her reply to The Question was that nothing would be different in her life, except that she wouldn't suffer the embarrassment of falling asleep so frequently and so abruptly. The disorder did not interfere with her functioning in any debilitating way, since, when she was active and engaged in taking care of the children, cleaning house, or doing any other physical work, she was almost always free of the symptom, and fully alert. Nor did the "misery" of her embarrassment restrict her from continuing to ride the bus whenever it was necessary.

Our clinical impression was that she suffered with narcolepsy, a hypothesis strengthened by the fact of the couple's being African-American, and therefore from a group at greater genetic risk for this disorder. We referred to a neurologist, who confirmed the diagnosis, prescribed the appropriate medication, and put an end to the couple's need for further counseling.

30. What do you expect will come out of our work together?

This discussion tells us something about the person's relative optimism or pessimism, a key index to therapy readiness. Some clients have already been to many therapists. In some cases it may be surmised moving from one to another enables them to maintain the symptom with a clean conscience. Therapy can be a sideshow in itself, one of many excuses or rationales for not getting on with the real business of life, and even a way of certifying a person's being inadequate or unprepared for those tasks. When being in therapy is part of a neurotic arrangement we can expect to be told something like, "When my therapy is over I'll be able to address the problem of _____."

> - A woman of thirty-five said, "I have a good marriage, but my husband wants a baby and I don't. I told him that I can't deal with it until I get my head straight. Can you help me?"

> - A parishioner with a host of problems went to see the priest in an urban parish. The man's wife had left him, and he had the sole responsibility for the care of two young children. He was about to be evicted from his apartment, and he was in danger of losing his job. He attributed all his difficulties to a conspiracy being directed against him by the Elks Club, the Presbyterian Church, and the Democratic Party. The priest recognized a paranoid arrangement, but felt incompetent to be of help to the man. Gently, he recommended that psychiatric help was needed. "You're absolutely right,

Father," replied the distraught parishioner, adding, "As soon as I can get these Elks, Presbyterians, and Democrats off my back, the first thing I'm going to do is go to a psychiatrist. This whole thing has shot my nerves."

In this case, pastoral counseling was his side show; by means of it he was evading a realistic reappraisal of his failed marriage, hoping to gain some measure of sympathy, and (by concocting such a dramatic explanation to account for his troubles) to salvage some scraps of self-esteem.

The Life Tasks: Love, Work, Friendship/Community (Items 31-36)

LOVE: Situation and Goals

31. Situation. Tell me about your love relationships. If the client lacks relationships of emotional or sexual closeness, ask: How do you account for this?

Prompting questions (these questions are relevant regardless of a client's sexual orientation): What makes a man "masculine" to you? What makes a woman "feminine" to you? How do you compare yourself to these assumptions for your sex? Do you experience difficulty expressing love and affection for others? Difficulty receiving such expressions from others? What does your partner complain about in you? What do you complain about in your partner? Describe your first encounter with your partner. What was there about him or her that impressed you at that time?

Note that we use the term "prompting questions" here. We prefer a free flow of conversation to follow out of the instruction, above, "Tell me about your love relationships," but sometimes the client does not know where to begin, or may overlook key issues. In such a case one or more of these questions may serve as a guide.

The questions about masculine men and feminine women are significant regardless of sexual orientation; they are designed to reveal the client's gender guiding lines in a brief sketch. Requiring little reflection, the replies express usually unexamined biases regarding the two sexes, naming qualities, competencies, and attitudes likely to be related (directly or antithetically) to the person's basic perceptions of Mother and Father. Here are a few examples, male and female, with ages noted in parentheses:

Male (30)

Masculine

The way you walk; tone of voice; way you handle yourself; a strong person, able to handle your own problems.

Feminine

Not loud; not aggressive; visual (how she's built); soft; warm; someone who can comfort you.

Male (53)

Masculine

Strength and endurance, both emotional and physical.

Feminine

Warm; empathic; a kind of teacher (my daughter taught me to play handball. I wouldn't have tried it if she hadn't conned me into it!).

Female (37)

Masculine

Gentleness; intelligence; humor; nurturing.

Feminine

I think of negatives: not girlish; not ruffles; not a tomboy either; not dumb; not up tight; laid back. (This is harder. I'm pretty clear about that "masculine man!")

Gay Male (36)

Masculine	*Feminine*
Provides whatever is needed, no matter what it takes you to do it; you can right any wrong, and whatever befalls, you can correct it.	Looks.

The prompting question, "How do you compare yourself to your assumptions for your sex?" brings forth a self-assessment:

- A man of fifty-six: "I stack up pretty well. I don't think my wife would agree with me, but I'm pretty comfortable with it."

- A woman of forty-two: "I am loving and giving. But I'm afraid to be flirtatious. I envy women who aren't afraid!"

- A lesbian of twenty-three: "I'm not like this at all. I'm not feminine in any traditional sense. Isn't it odd that I'd have a list like that?"

- A man of forty: "I don't do very well with the 'toughness' and being able to handle situations. I'm O.K. on the 'caring' and 'sensitivity.'"

Here are examples of responses to the questions, "Do you experience difficulty expressing love and affection for others? Difficulty receiving such expressions from others?"

- A man of forty-eight: "I can really dish it out. I treat my girlfriend like a princess. I hug her a lot. But I don't want her to move in on me. Actually, she's sort of unhappy about this. I don't even like her to give me presents."

- A woman of thirty-two: "My mother says I'm thoughtless, that I don't show her enough attention. She's not a warm person. She's so defensive. With other people it's easier for me, but still a problem. I like it when they're warm with me, but I'm still a little awkward in the way I respond."

The next two questions "What does your partner complain about in you? What do you complain about in your partner?" can elicit further information about the client's masculine and feminine guiding lines, and, regardless of orientation, the client's training for intimate cooperation.

- A man of forty-four: "She says I don't communicate with her. As she puts it, 'Once I cross that line with you, it's as if I don't exist.' She says I'm not affectionate enough. I tell her she talks too much. Also, she's quick to anger. She flares up. But once the anger is out, it's forgotten. I can stonewall, and I bear grudges. I wish I were more like her."

- A woman of thirty-seven: "My boyfriend doesn't like it that I'm a Democrat, and that I hate to cook. He complains about my temper too. I blow my fuses, especially when he interrupts me, which he does a lot. Then he complains that I don't speak up calmly and directly. So, I don't like the interrupting, and I don't like that he smokes and that he's married. I complain about that a lot."

- A woman of twenty-eight: "His mess drives me crazy. He's a pack rat. Also, he's overcommitted, tries to do too much, and is never home. He doesn't like the way I try to organize his life. He says, 'Leave me alone.' But I can't. Organization is important to me. He also doesn't like the way I worry about him. But he's out on the road, in all kinds of weather! How can I not worry about him?"

The last two of the prompting items, "Describe your first encounter with your partner. What was there about him or her that impressed you at that time?" are directed by the concept of the

FECK or, "First Encounter of the Close Kind" (Belove, 1980). The character of the relationship can often be discerned in the story of the first encounter, which reveals the qualities that attracted the two to each other. Their movement in the story may also serve as a paradigm of the ensuing transaction between them. They may continue to rehearse this transaction successfully, or it may have failed them in meeting the challenges of the current situation, leaving one or both of them discouraged.

Research (Willis and Todorov, 2006) tells us that a momentary (100 milliseconds) face-to-face encounter with another allows for judgments to be made of the traits of attractiveness, likeability, trustworthiness, competence, and aggressiveness, and that these judgments survive over time, often increasing in confidence. The authors' findings "suggest that minimal exposure to faces is sufficient for people to form trait impressions, and that additional exposure can simply boost confidence in these impressions" (p. 597). Here is an exemplary case:

> - A man of thirty-six: "I met my wife at the laundromat. We only spoke casually. She had beautiful eyes, and was attractive. She seemed to be a professional, or at least someone who used the language correctly! Also, I liked the fact that she was cautious. She didn't want to carry on an extended conversation with a stranger, and wouldn't tell me where she lived. That was feminine. I asked around, and tracked her down. Finally we went out, and it turned out that we liked a lot of the same things. She has always controlled the relationship, though; her cautiousness again, I guess. It keeps me in line."

In the case below, however, the client took a *second* look and made a *second* evaluation, and the relationship is troubled:

> - A woman of forty-one: "I didn't like him when I first met him. But he was very persistent, kept calling me up and asking me out. I resisted until I saw him again at a party. There he was, surrounded by a lot of friends. I saw that he

had a strong personality, and was the leader of the group. He was attractive and outgoing. Before, I had thought that he was too aggressive. He paid a lot of attention to me that night, which I liked. He's too busy now."

32. <u>Goals</u>. What do you want to improve or change in this area of your life?

In eliciting an ambition to improve or change the present situation, we convey that improvement and change is possible, emphasize that life is purposive, and get hints of clients' privately perceived stumbling blocks. On the other hand, the reply may indicate this is an area in which the person is experiencing success; therefore talking about it can be a source of encouragement.

- A woman of fifty: "I don't know if I want my lover to come back or not. I have so much resentment. She's hurt me so much. If she could be like she used to be, then I'd want her."

- A woman of twenty-four: "I want to get married, but my boyfriend doesn't. He says he thinks things are just fine the way they are with our living together. But to me it's just playing house. I want a real house, and children. How can I get him to change?"

- A man of thirty-four: "I'm happy with this part of my life. I feel happily married. My wife and I have a lot of fun. What worries me is the job situation."

<u>WORK: Situation and Goals</u>

33. <u>Situation</u>. Tell me about your work. If the client has no occupation or is currently unemployed, ask: How do you account for this? Explore interests and ambitions in this area.

Prompting questions: What has been most satisfying to you in the jobs you have held? Least satisfying? What other work have you done? Why have you left those jobs? Are you aware of anything about the way you work that causes trouble for you (e.g., procrastination)? Do you feel appreciated at work? How do you evaluate yourself in relationships with others at these levels: superiors, peers, subordinates? Those of the other sex or a different sexual orientation? Of a different ethnic/racial group?

The client's perception of areas of personal effectiveness and ineffectiveness, and an evaluation of these, are invited by the first of the prompting questions, "What has been most satisfying to you in the jobs you have held? Least satisfying?"

- A woman of forty-four: "I like the technical part of being a nurse. I didn't like being a floor nurse who took care of people. I liked working with the doctors in intensive care where I had a lot of medical responsibilities. I was very competent in those areas. On the floor they complained that I didn't work fast enough. They were pushing patients through there like cattle, and we were really overworked. I never felt like I did a good job. I liked the feeling of being a member of the team in intensive care, and the doctors liked my work."

- A man of thirty-six: "My present job is blah. There isn't anything new for me to learn. The challenge is gone, and I'm getting stale. I'm not afraid of responsibility, and I make good decisions. They respect me. But if it isn't exciting, it just doesn't interest me. I don't know how much longer I can keep it up."

In the prompting questions, "What other work have you done? Why have you left those jobs?" we are looking for a pattern of experience and attitude.

- A man of sixty-three: "I've done a lot of different things. I've worked since I was fourteen. Summer jobs in a grocery, in a factory, you name it. When I've left, it's been to move on and move up. In college I got into business administration courses, and realized that was what interested me. I'm not interested in the product. I can work in any industry as long as the pay is good and I can live the kind of life I want."

Self-evaluation is invited by the question, "Are you aware of anything about the way you work that causes trouble for you?"

- A woman of fifty: "I don't delegate. We're under a lot of pressure. It's a lean, mean company. This has become more and more of a problem for me."

- A man of forty-seven: "I'm disorganized. I don't stick with a job until it's done. I get distracted by my thoughts about other things. I leave the paper work, and make phone calls; or I stop making phone calls, and go to the paperwork. I'm my own best interrupter."

- A man of twenty-nine: "I don't do the follow-up work. I love new projects. Once they're in place I lose interest in them, and depend on others to do the detail work — which they don't like. I get complaints about this."

- A man of fifty-six: "I procrastinate. The Bar Association has brought two complaints against me, based on my client's reports to them. I do good work, but I can't seem to bring it to completion in an appropriate length of time."

Biases regarding unfairness, discrimination, or prejudice may be revealed in response to the next question, "Do you feel appreciated at work?"

- A man of fifty-three: "No. I move too fast for them. If they could fire me, I believe they would. Luckily for me, my immediate boss has a lot of clout. As long as he's there I'm untouchable. I like making decisions and acting. I don't like waiting for authorizations from upstairs. When I can't get decisions pronto, I just go ahead. So, even though I'm generally unappreciated, the person who counts appreciates me."

- A woman of thirty-two: "My peers are jealous because of my success, and they aren't very nice to me. This troubles me. I'd like to be a team member. As it is, I'm a loner. I'm a very successful salesperson, though, so I have that going for me. The others think I'm a snob. When I try to be friendly and helpful, they think I'm faking."

The next prompting question elicits a self-assessment. "How do you evaluate yourself in relationships with others at these levels: superiors, peers, subordinates? Those of the other sex or a different sexual orientation? Of a different ethnic/racial group?"

- A woman of forty-two:

Superiors:	"Good, cooperative, mutually satisfying. I work best when I get encouragement, good feedback, and have the freedom to set my own objectives."
Peers:	"It's a team effort. I think we respect each other's areas of expertise."
Subordinates:	"I expect them to do a good job. And I like to see them succeed. I think they work well with me."
Other sex:	"I make it a point to get along with the men in the company. I'm careful not to play cute or invite flirtation."
Sexual orientation different from yours:	"As far as I know, there aren't any gays at my level, but there are a couple of lesbians. I'm ill-at-ease with them. If someone made a move on me, it would really be awkward. I don't know how I'd handle it."

Different ethnic/racial group:	"It's important to me to treat people well, so I kind of go out of my way for people if I think they're at a disadvantage — and some people are — especially if they're black."

- A man of thirty-seven:

Superiors:	"Fine."
Peers:	"Not so good. I'm quite a bit above them in educational level, so I don't enjoy them. They hang out together socially, and I don't take part. I don't think they use their brains."
Subordinates:	"Good."
Other sex:	"There are very few of them in the industry. I feel kind of sorry for the ones who are. They just don't fit in. I try to help them out."
Sexual orientation different from yours:	"It's O.K. They stick together and that's fine. In my company, we're expected to be polite, and we are."
Different ethnic/racial group:	"Yeah. Again, I'm O.K. with it."

34. <u>Goals</u>. What do you want to improve or change in this area of your life?

> - A man of forty-one: "I wouldn't want to change anything. I really like my situation; I can see it growing and getting even better in the years ahead."

> - A woman of thirty-five: "I want to quit my job and go back to school. I like working in the travel business, but I know that I can't move as far up as I want to unless I have more education. I thought this was one business where not going to college wouldn't matter, but I've learned that I just have to know more about everything."

> - A gay man of thirty-three: "When I finish law school, I'd like to specialize in family law. There's a real need for sensitive men in this field."

> -A lesbian of twenty-eight: "I'm over-qualified for my job and they know it. I should be promoted. I think my sexual orientation holds me back, but I can't prove it."

FRIENDSHIP AND COMMUNITY: Situation and Goals

35. <u>Situation</u>. Tell me about your friends and your life in the community. If the client reports few or no connections with others, ask: How do you account for this?

> *Prompting questions*: From where do you draw your friends? How many close friends do you have? How often are you together? What do you do? Do you have friends of the other sex? Different sexual orientations? Different ethnic/racial backgrounds? How do your friendships end? How much do you feel you are able to confide in your friends? What sort

of impression do you think you make on people the first time they meet you? Does this impression change over time? If I called a close friend of yours and asked, "What do you value in [client's name]?" what would he or she say? What kind of connections do you have in the broader community (e.g., volunteer work, athletics)?

By the question, "From where do you draw your friends?" we inquire into the extent of the client's community feeling. It is telling if friends all derive from the family of origin and the extended family, or from members of the ethnic, racial, or religious group in which the client grew up. When the loyalties and affiliation of a person are not extended into the wider community, we can expect to see an increased intensity of symptoms and disturbances. *The extent of community feeling and the severity of the dysfunction are in an inverse proportion.*

- A woman of thirty-nine reported that she had no friends. She remains embedded in her ethnic (Turkish) family. Her social life was limited to her sister, her brother-in-law, nieces and nephews, and her widowed mother. There was no man in her life, and she described her colleagues at work as "competitive and unfriendly."

- A man of twenty-seven reported that he drew all of his friends from his circle at work. "We're all techies. We go out drinking and talk computers. The group includes a couple of really bright women. We keep sex out of it."

- A man of fifty-one reported, "One of my gripes is that my wife doesn't make her own friends. She gets mad at me because I don't always want to include her in my evenings 'on the town' with my buddies. I tell her to get some friends of her own. My friends come from work, from my health club, and from my connections with United Way and other activities in the community."

A further question with respect to community feeling is, "How often are you together? What do you do when you are together?"

- A woman of twenty-six: "My best friend is a sorority sister from college. We have lunch together at least once a week. I tell her my troubles, and she tells me hers. I stay in touch with a girlfriend from high school. It's been hard to make new friends in the city. I don't like the bar scene, and I don't know what else to do to meet people. People at work aren't suitable. They don't share my interest in the arts."

- A man of thirty-one: "I don't have a big group of friends, but I'm happy with that. I was always the type who had one 'best friend.' So I prefer a small group. We do lots of things: sailing, skiing, tennis."

Also, "How much do you feel you are able to confide in your friends?"

- A man of thirty-eight: "I can tell Andy and Joe anything. They know all about my divorce and the troubles I'm having. They give me good advice."

- A man of twenty-three: "I've always been kind of an introvert. In the past, it's been hard for me to open up with *anyone*, but since I came out and got involved in the gay community, for the first time, I not only have friends, I have a partner, so I've learned how to talk about things I've kept private for years. It's been a revelation!"

- A woman of thirty-two: "I keep them at a distance. I like to keep it light. When I have trouble I talk to my father or my brother. I don't want my friends to know about it."

Further, "What sort of impression do you think you make on people the first time they meet you? Does this impression change over time?"

Most clients have not thought about these questions before, and are intrigued to think about them. Some "have no idea," which may be an index to how little social effectiveness they experience. The degree of self-confidence and self-understanding may be revealed.

- A woman of forty-eight: "People see me as friendly. I'm a good listener, and I'm honest. I don't talk much, but when I do it's because I have something worthwhile to say. Most of my friends are black, and we're very tight. I keep my friends a long time. I give them a lot of attention, and they appreciate it. They can rely on me. They know I'm there for them when they need me."

- A woman of twenty-five: "They think I'm aloof and maybe a little too quiet. I like to feel my way into a situation, sort of size it up before I get too involved. I think they see me as attractive and intelligent. After they get to know me they see my sense of humor, my talent, my warmth and caring. I think some are threatened by me, especially men."

- A man of thirty-six: "I know what I want them to think, but I don't know what they think. I want them to think I'm intelligent, articulate, suave, and put-together. The only thing I'm sure of is that they see me as tall!"

And, "If I were to call one of your close friends and ask, 'What do you value in [client's first name]?' what would he or she say?"

- A man of twenty-two: "They'd say that I'm loyal, that I care, and that I'm supportive."

- A woman of twenty-eight: "I don't know what they'd say. Well, I give them good advice. They know I'm sincere no matter what I tell them, so they can count on me."

- A man of fifty-three: "They know they can rely on my support, whether I agree with them or not. So I guess it would be my bluntness that they would value most."

Finally, "What kind of connections do you have in your community?"

- A woman of sixty-seven: "I've lived in this town all my life. At one time or another, I've been involved in the church, in the schools, in different charity activities, on the library board, and so forth. I've given that up now. It's my turn! Now I want to go to Florida."

- A woman of thirty-five: "I'm active in the women's work of the synagogue. I feel that I have an obligation to do that as long as my children are involved. I don't especially enjoy it, but what can I do? I feel this way about a lot of things in my life: I *ought* to do them, so I do."

- A man of forty-one: "I'm busy with my church. I'm on the vestry. It's important to me. I used to be a Catholic, but I'm gay and not accepted there, so it's like I've found a new home, and I'm glad to feel 'at home' in it. I also do volunteer work with a group that takes care of old people. I keep thinking about what it will be like for me when I get old, since I have no family. I hope this group will still be around then."

- A man of twenty-four: "I don't have any connections in the community. My parents know everyone and are real active. I'm just going to school. I can't see ending up like them. I want to get away and travel."

36. <u>Goals</u>. What do you want to improve or change in this area of your life?

- A man of twenty-nine: "I'd like to be a better friend, more consistent, less volatile. I'd like to be very consistent in my conduct, and get balance in my life — social, marital, professional, religious."

- A man of thirty-four: "I'd like more and closer friends, and a more active social life. I'd like to feel comfortable among strangers. I used to, back at home . . . in the Czech Republic, when I was young. Then I came to the U.S. to go to school, and never really went back. It's never been the same for me."

- A woman of thirty-two: "I want to feel that I can say what I want. Other people seem to feel free to do that, and it always surprises me! I just kind of go along with whatever is happening. Then I get resentful."

Conclusion

In this chapter we have presented the *IPCW* format for the initial interview inquiry, along with samples of responses elicited by each of its numbered items.

In the next chapter we move on to the inquiry process in the lifestyle assessment, with a detailed consideration of the way information about the client's position in the childhood family can be gathered, and with a guide to eliciting and recording the client's early recollections.

While this sequence fits with the way the text is organized, a reader may not want to consider gathering more information without seeing how we use the data. Skip to Chapter 5, "Life Situation, Presenting Problem, and Life Tasks," to see how we make sense of it as we review everything with the client: It's here that we test our understanding and our interpretations. Note that the final step in the initial interview is composing a summary of the material; the method for this procedure is in Chapter 12.

Chapter 4.   LIFESTYLE ASSESSMENT INQUIRY

We divide the lifestyle assessment inquiry into two sections. First we systematically record a thorough description of the members of the person's family of origin, their relationships, and their circumstances. This gives us a sense of the social dynamics and organization (the "family constellation") in which the person first encountered the tasks of life. With this as a context for understanding, we then elicit and record memories of specific incidents that occurred in the early childhood situation ("early recollections") to gain a picture of the evaluations the person made (and *makes*) of life.

The form we use for the investigation into the family constellation has two parts: "Part I: Parents, Other Adults, Milieu"; and "Part II: The Situation of the Child."

The form for recording early recollections provides space for eight stories of childhood experience, plus, for reasons explained below, a ninth recollection. (For the complete outline of the lifestyle assessment inquiry, see *IPCW*, pp. 26–32.)

This is a formal way of structuring the work that has the advantage for the therapist of keeping the discussions with a client disciplined and to the point, and the advantage for the client of providing a reassuring sense of direction, welcome as an antidote to the disorientation that occasioned the consultation. Even so, we remind readers that a living conversation is never so neat as to fit a formal structure, so the recording therapist must remain alert and nimble.

A client's memory is likely to be stimulated during the family constellation inquiry so that some early childhood recollections (ERs) will come to mind then, prior to the place in the outline at which you intend to ask for them. Each of these recollections should be recorded at the time it is presented, on a separate sheet of paper which can be interleaved with the other data as a kind of footnote.

Because these recollections are prompted by the material under discussion, they do not have the same projective quality or value as those memories that the client presents without an awareness of any context or reminders. They are important, however, and we take them into account when we compose the *Summary of the Family Constellation* (as you will note in the following case).

Later, when you are making the formal record of early recollections (unprompted by anything other than questions such as, "How far back can you remember?" and, "What is the next memory that comes to mind?"), the client may recall some of these same incidents again. If so, make a fresh record of these memories, encouraging the client to tell the story once more as he or she is now remembering it.

This chapter includes a verbatim transcript of the lifestyle assessment inquiry in "The Case of Dan." We have protected confidentiality by disguising all identifying material, as we have also in the other cases referred to in this book. To introduce "The Case of Dan," the lifestyle inquiry transcript is preceded by a synopsis of the material recorded in Dan's initial interview and followed by the *Summary of Impressions Derived from the Initial Interview* (see Chapter 12 for the method for composing this summary). Concluding this chapter are three summaries derived from Dan's lifestyle data. These three summaries are presented here without discussion of the methods used in shaping them, which are covered extensively in Chapters 13 and 14 of the text.

The Case of Dan

General Diagnosis: The Initial Interview

[Note: The following is a synopsis of data taken from the record of the initial interview inquiry in "The Case of Dan."]

Dan was forty-five years old. He had been separated for eight months from his wife, Eileen, their fourth separation in fourteen years of marriage. Before marrying they lived together for two years. He and Eileen had a daughter eleven years old. He was stepfather to Eileen's older daughter from a prior marriage.

The disturbance between him and Eileen was the presenting problem. He said they loved each other, "at least we say we do." Their relationship was stormy. He reported he had acted "violently." When asked about it, he said he had never struck her or in any other way physically abused her. He had "raged and sworn" and struck his head against the wall in "displays."

Dan had one sibling, a brother three years younger, who was married and had no children. Douglas was a violinist with a metropolitan symphony orchestra.

Dan's original family was intact. His mother was seventy-three, his father, seventy-six. They still lived in the original family home in Virginia: "They went through some rough periods, but stayed together, and are very happy and contented now. It's funny, because when I was a boy I was always afraid she would leave him. I felt sorry for her, somehow. She was very ambitious, and Dad certainly was not, though he was terribly bright. She was always disappointed. When I look back on it now I can see that he was a hard worker, and a good provider, but it was never good enough for her. I never saw my father get angry with anyone but her. I guess her complaining got to him." Father was described as arthritic, but "in good spirits"; Mother, as having been troubled for twenty years by a smoker's cough, but "she refuses to go to a doctor, and is very active physically."

He assessed his own health as "generally good, though sometimes my back goes out. I can stand the pain, but I can't stand the immobility." He does not report any trouble arising from the use of alcohol, drugs, or food, and is not under a doctor's care.

He had been in therapy with another psychologist for six months, and decided to try someone else because "I want the process accelerated." When asked about the outcome of the therapy, he said, "He helped me to define some problems." Dan said that he had described himself to the other psychologist, "with a smile," as "indecisive" and told him, "I'm a person who likes to be told what to do. Either help me to become more decisive, or tell me what to do." Asked about his present goal in therapy, Dan replied, "I want a better feeling about myself and a better understanding of why my relationship with Eileen is so difficult. I want an answer to the question, 'Can we make a happy marriage?' I want to be better at making decisions, being understanding, and sharing affection."

Dan said the trouble between him and Eileen had started the second year after they met, before they began living together. "I was already aware that we were going to have some struggles. When we met she was still married to her first husband. I was doing well and was happy and carefree. She was very unhappy. Her husband was an elitist, a rather learned man, from a family that was socially 'above' hers. He was condescending toward her, more a mentor than a husband. It hurt me to see it. She struck me as a remarkably bright and 'gutsy' sort of person who deserved better. I was always very shy with women, and never thought of myself as attractive to them, at least not to the ones I saw as attractive. She had striking good looks, and I was impressed by how friendly she was toward me. Her husband's coldness, and before that her father's indifference to her, had left her with some pretty bad feelings about herself. She had the idea she was unlovable. I knew better, because I had fallen in love with her."

He reported that she left her husband for him, "and then, almost immediately, she became very insecure, and accused me of infatuation, of being immature, incapable of true love. It was always, "If you love me, prove it, show me that you love me."

He said that he had "put a lot of effort" into this task, had given up some attractive job offers, and had moved to another state "where she had always wanted to live. But it was never enough.

She says I'm selfish, that I don't recognize her emotional needs, and that I don't share my feelings. Well, I guess I am rather selfish, and I do have a hard time trying to guess what she wants from one day to the next. I don't feel able to share my feelings with her, except for my anger."

Work.   Dan reported that he was a metallurgist, currently manager of a large research project. He liked working with people from a wide variety of backgrounds. "I like synthesizing something out of nothing in my work. I like organizing a team out of disparate pieces, bringing order out of chaos. Some people philosophize forever; I like to get things done." He then added, "Eileen has played a heavy hand in the course of my career. She made the decisions about where we lived and my jobs."

About working with his peers, he said, "I felt competent until my present job. Most of the men are recent Ph.D.s, and have lists of publications as long as your arm. They regard themselves as an intellectual elite." He reported that he got along well with subordinates, but added, "I'm a little afraid of supervising them. I don't find it easy to force people to do things, though I can organize teams on a project."

He said that he worked well alone, and avoided supervision. "I can never tell where my boss is at. He's hard to read. He always knows more, or at least appears to. He's intellectually intimidating to me. I think he'd say of me that I'm a 'soft' person, which isn't exactly a compliment among 'hard' scientists."

His goal in this area: "I don't know what I want to do. I love woodworking and puttering around, and I love music. Maybe I should go into furniture design. It's ironic, though: I got my Ph.D. so I could be secure in a job and able to do whatever I wanted to do. Now, it's as though I'm stuck with metallurgy because of it."

Love. Dan said that he had trouble expressing affection for women "except my daughter. I'm very affectionate with her." He reported impotence, not only with Eileen, but with other women as

well. "I can't get close to people. I don't know what to talk about."
His first experience of impotence "was when Eileen forced me to
make love to her because she wanted to get pregnant. She said we
had to have a child. She had had a couple of abortions before that,
and I had begun to draw away from her, especially after the second
abortion. But I love Eileen. I wish our marriage could work out."

Friendship and community connections. Dan reported that he
had a few close friends, none of whom lived nearby, even though
he had lived in the same neighborhood and worked at the same job
for two years. "I've begun to change that in the past few months,
since Eileen and I have been apart." He had good friends in other
states, but saw them only occasionally. His best friend was a
woman. They talked on the phone "a great deal" and have
considered living together. His next best friend was also a woman
with whom he had had a "periodic" love affair. "It was a 'same
time, next year' type of thing. I've drawn away from her now.
She's a real 'earth mother' sort of figure. My other friend is a new,
modern woman — very clear, direct, ambitious. We're both
opposed to marriage, but we're considering living together or at
least in the same community. We'd like to be close."

Of his male friends, he said, "It's mostly casual. You know, a
drink after work. I've always felt shy around people, especially
girls and women. It's always been a problem for me. Eileen says
that it's my emotional immaturity and self-centeredness. The
friendships I do have don't end though; they just change. Eileen
faults me for that, too. She says, 'You're forever looking someone
up; you live in the past.' I suppose she's right. It may even be part
of my trouble with Eileen; we've been separated four times, but we
never end it."

He believed that his friends valued him for being easy to talk to,
understanding, sympathetic, and sensitive. On first meeting, "They
see me as open, friendly, and easygoing. It doesn't always hold up.
I reach a point where I can't be supportive, where I switch off. It's
because I'm afraid they're going to demand something I can't cope
with. In a way, I keep people at a distance. I need to be thought

well of, and if they really knew me, they'd find out that I'm not as sensitive as I'd like to appear. I can be pretty selfish, too."

His goal in this area was to have a wider circle of friends. "I've been going hammer and tongs with Eileen, and haven't had time. In the last six months or so I've realized that other people are more important to me than I used to think they were."

The Case of Dan (continued)

Summary of Impressions Derived from the Initial Interview

[Note: Immediately after reviewing the above data with the client and the recording therapist who reported the material, the consulting therapist dictated the following *Summary of Impressions*.]

Dan is emerging from a great defeat, reluctantly separating himself from an ambition which promised everything if it could be attained, but which demanded that he give more than everything he had in exchange, and, when he had given it, still proved unattainable.

The ambition appears to be based on a misunderstanding in which Dan long ago pictured manhood as an *achievement*, and was consequently unable to enjoy it as a *given*.

In the course of his childhood training the questions of sexual identity, personal power, domination, and submission seem to have been very much in the forefront, possibly because of Mother's struggle to demonstrate her importance, and her struggle to have Father work harder in order to insure it.

It appears there was a general atmosphere of struggle in Dan's childhood family, with Mother energetically pushing, and Father trying to resist and comply at the same time. Further investigation may uncover the lines of this struggle in detail.

Dan probably wanted to be a good and diligent man like Father, but did not want to be in the service of a demanding, managerial woman like Mother.

As the firstborn son he might have disarmed Mother by working to remain her favorite and to become her protégé, learning from her how to manipulate, and then manipulating her so as to gain the advantages of her protection and pampering.

With other women it was a different matter from adolescence on. What he hoped for with them was to emerge as the Real Man that Father had appeared unable to be. This would require that he combine intellect with striving for accomplishment, and do so in such a way as to demonstrate a superior "goodness" which would mark him as "sensitive" to the "needs" of the woman. Then he would be able to rule without having to fight for domination.

He would be all things to a bright, gutsy, but unappreciated and unhappy woman, making her happy, and making it unnecessary for her to complain ever again. Thereby he would demonstrate to his own satisfaction that he had solved the problem of manhood by making an alliance with a powerful (and powerfully resentful) woman who would look up to him, rely upon him, be grateful for his superiority, and give up her resentments by yielding to the argument of his "goodness."

The problem that he hadn't foreseen, however, lay in the fact that the more superior, the more admirable, and the more able he looked, the more would be demanded of him, the more he would be required to be endlessly and generously concerned to make it up to the inferior woman for her disadvantages, and (the trickiest part of all) to prove to her that he was worth looking up to by never looking down on her.

Until now, his way of dealing with the unsolvable problem he had set for himself (that is, the problem of making the Unhappy Woman happy) was to work harder at finding a solution. Only now is he ready to acknowledge that all the lore of the metallurgist is still unequal to realizing the alchemist's dream of transmutation: Lead remains lead, and gold remains gold, and gold (like a woman prepared to be happy with him and with life) is where you find it.

The Case of Dan (continued)

Special Diagnosis: The Lifestyle Assessment

[Note: The following is a verbatim record of the inquiry in the lifestyle assessment in "The Case of Dan," following the form of the *IPCW*, Items 1 - 33.]

Client's present age _____45_____

Date __XX_____

_____

_____

Family Constellation, Part I: Parents, Other Adults, Milieu

1. Father's name ___Ronald_____

   Age at client's birth _31_ Age, if living _76_ Or, at death __

   Year and cause of death _____

   Client's age at Father's death _____

2. What kind of man was your father when you were a child, up to age 10 or 11, in the pre-school and grammar school years? Consider occupation, activities, personality, health, level of education, and values (i.e., what was important to Father).

He always wanted a daughter. My brother and I both knew it. He even had a name for her: "Evelyn." He didn't take care of himself. Mother waited on him. He was quiet. He seemed very capable in a quiet sort of way. He wasn't able to go into the army — bad feet — but he was in the National Guard or something during World War II. He was a warehouseman. He always seemed to be there to take care of us, my brother and me. Dad sang hymns. I remember his singing hymns with me sitting on his lap. He kept a "victory garden," and he used to take me out back to work in the garden with him.

Dad never hurried. Mother would shout at him, "Hurry up or you'll miss the bus." He'd say, "Never mind. There'll be another one." He used to take me to see his father, my grandpa, who'd always be sitting in his chair in the kitchen with a cat on his lap. My grandma and great aunt were both paralyzed — my great aunt was born with some sort of "creeping paralysis," as they called it then. Grandpa looked after both of them.

My dad flirted with women, but there was never any indication that he was unfaithful. Dad's really a good, earthy, sensible human being. Very little ever bothered him, except my mother. Mother said that he wasn't very good with sick people, but I could never understand what that meant, since she was never sick, or at least never admitted it if she was, and he always seemed to me to be kind to his mother and my great aunt.

Dad was self-made. I don't know if he finished eighth grade. He learned fast, and was really a genius at figures. He was always very healthy as a young man. He was a good worker.

3. How did he relate to you and the other children? Consider favorite child; discipline and how you felt about it; expectations regarding your behavior and achievement.

I think Douglas was his favorite. Maybe this was a reaction to Mother's favoritism toward me. I didn't really perceive this in the early years, though; I began to see it when I was a teenager. He'd discipline us with verbal threats and occasional smacks, but he was never cruel. Mostly it was just by his presence and authority. We weren't afraid of him. I don't have any idea what he wanted me to do with my life. There just was no indication. And he was fairly liberal about my behavior. Not much bothered him.

4. How do you see yourself now as like Father? Unlike Father?

I'm like him in that I don't expect too much of people. For example, I've guided my children; I've never dictated to them. Also, I like working with my hands, and I like cats. I think I'm easy to talk to — as he is now. I'm different from him in many ways. I'm achievement-oriented. I'm much more energetic — going places, doing things. I will hurry, and I will worry. I went on to higher education. He never regarded education as important. He said, "I left school at fourteen. That was good enough for me."

5. Tell me briefly about Father's background and family of origin.

Dad's family is very diffuse. His brothers and sisters weren't close. They got into fights and arguments. When I was eight or nine Dad and two of his brothers had a scene at my grandmother's funeral, and he hasn't talked to them since. Mother said Grandpa spoiled and idolized Dad. There were seven or eight children. I think that there were two girls and five boys, and that Dad was one of the youngest. My grandpa was a gardener. I know they were very poor, and that they all left school quite young. I told you that my grandma wasn't well. Yet she raised those kids.

Mother said that my grandma was treated very badly by Grandpa, but I didn't see that. My mother was always saying to me, "For God's sake, don't grow up to be like an Allen!" In other words, as she saw them, don't grow up to be bad-tempered, inconsiderate, and someone who can't stand sick people.

6. Mother's name   <u>Edith (Edie)</u>

   Age at client's birth <u>29</u> Age, if living <u>73</u> Or, at death ___

   Year and cause of death _____

   Client's age at Mother's death _____

7. What kind of woman was your mother when you were a child, up to age 10 or 11, in the pre-school and grammar school years?

Consider occupation, activities, personality, health, level of education, and values (i.e., what was important to Mother).

She was a housewife, energetic, busy as a bee, hard-working, quick, kept a nice house, cooked good meals. She was very ambitious for me. She hung over me when I was growing up. It was "nose to the grindstone" for me. She encouraged me to work hard — really beyond my ability. I remember that I got real low scores on the SATs. She made me take them again, and I did better. But that was later. She was determined that I wasn't going to be deprived of the opportunities that came with higher education. I also remember her getting very angry with Dad.

Early Recollection: I was three or four years old. I was sitting on Dad's lap. He's singing to me or telling me a story. Mother walked in holding a carving knife and said, "I could cut your throat!" He says something and she shouts at him and leaves the room.

Most vivid moment: The look of hatred on her face when
                    she says it.

Feeling: Frightened.

She demanded that I look spotless, well-groomed, and that I was perfectly behaved. She sewed and ironed. I wasn't supposed to get dirty. I had to be careful. At age six or seven, I had a couple of baby teeth knocked out. Later, I broke the bottoms off two permanent teeth and she screamed at me, "You'll be disfigured for life!" She was horribly upset. It was a mortal blow! She wasn't concerned about the pain I was in!

Early Recollection: I was eight. I woke up with a crick in my neck. Mother took me shopping. She kept telling me to straighten my neck. She got very angry. She said, "I'm ashamed of you. You're stupid, dumb, an idiot," or words to that effect.

Most vivid moment: Her trying to force me to straighten my
neck up, and I really couldn't.

Feeling: This is unfair! I felt very self-conscious and
ashamed. I also felt inadequate: I felt I *should* be
able to do it.

She treated me like a showpiece and protected me against my
will. She'd intervene with my teachers, for instance. I felt pressure
all the time, especially from the time I entered school.

She got colds very easily, but she'd "carry on." She was a martyr.
She stayed up and suffered. She had three or four miscarriages
when I was young. Even at these times she'd be on her feet. She'd
say, "I've got to feed your father and you. Your father's useless,
and you're a bungler."

My mother hadn't been able to finish high school, much less go
to college, and she always resented it. There wasn't enough money
in her family for that, not in those days. She was very musical,
played the piano, and sang. But she never could afford a piano.
She felt she'd married beneath her when she married Dad. They
were both of English ancestry; but she always said her people were
of "better stock" than the Allens because they'd been here since
colonial times, and the Allens hadn't come over until 1885. I don't
know if there's any truth to any of that.

8. How did she relate to you and the other children? Consider
   favorite child; discipline and how you felt about it; expectations
   regarding your behavior and achievement.

She was really focused on me. Douglas had an easier time of it.
She always claimed that Douglas and I had equal attention, but we
knew it wasn't true. She had a lot of expectations: You should look
good, be well-educated, dress well, be in the top ten of your class,
be well thought of by *everybody*. I stayed in the middle — I was
too talkative in school. She always said I could do better. She'd go
through periods after my report cards when she wouldn't talk to me

for a few days. I felt cast out and uncomfortable, and it really worried me. Oh! — and I had to have good manners. I was in the spotlight — I mean her spotlight — always.

Early Recollection: I was nine or ten. I'd been trying to make a little wooden house for my cat, and Mother said, "You're a bungler — a jack of all trades and master of none."   I remember walking into the kitchen and telling her that I was having trouble making the house. That's when she said that. Dad had only chuckled at all the nails I'd used.

Most vivid moment: When she said it to me.

Feeling: Inadequate. "I can't do anything right." That was the thought I had about myself. Awful.

As I said, if I did anything she didn't like, she'd withdraw her love and affection. She was always drumming into me, "Don't grow up to be an Allen." She'd say, "You're acting like an Allen." And, "I never did want to marry your father. He made me. He kept pestering me. When you kids grow up I'm going to leave your father." Whenever she said that I'd be scared for her. She'd be alone and friendless in the world.

9. How do you see yourself now as like Mother? Unlike Mother?

I'm like her in that I'm energetic. And for long periods I can be patient and uncomplaining. And I'm ambitious — not so much in terms of career and making a lot of money, but in terms of being thought to be proper, intelligent, informed, well-read. I'm not like her in terms of endurance. I can't endure what she endures. In the end I'll kick back. She always implied that she stayed in it "for the children." I couldn't do that. I am more tolerant than she is, but not as giving. A line I always heard was, "Your mother does anything for anybody."

10. Tell me briefly about Mother's background and family of origin.

There were more than ten children in her family. She was near the bottom, the second or third youngest. The one next younger to her died when she was in her twenties. He died of TB, and she stayed with him and took care of him during the illness. She said, "He was everything a man should be: sensitive, loving, and kind." She came from a very close family. They all looked out for each other. I knew her mother better than any of my other grandparents, perhaps because she was the only one who came over to our house. She died when I was nine or ten years old. I remember her as kind, gentle, and soft-spoken, and in the background, barely a real presence of any kind.

I didn't know my grandfather. I guess I saw him a few times, but I don't remember him. My mother never talked about her father. He was a carpenter who was in and out of work, and drunk as often as he was sober. He ruled with an iron hand. Three of Mother's brothers and sisters still live together. They never married. I can't really be sure about Dad's family; he says they're English, but Mother's is English through and through.

11. Parents' relationship. Tell me about how your parents got along with each other when you were a child. Did they express love, affection toward each other? Was one of them "the boss"? If they argued, what about? Who seemed to initiate the trouble, and what was the outcome? Did you take sides, openly or covertly? How did you feel about the trouble? Did you feel sorry for one of them?

They had a very stormy relationship. I always felt insecure. I was scared of the fights and of my mother's temper. Dad would shout back at her. There was an awful lot of shouting and insults and swearing. They never touched that I ever saw. They were never affectionate in any way. In fact, Mother wasn't affectionate with anyone. She hardly ever touched me. She would give me a little peck every now and then. That's all. I tend to think I sided with my father, but mostly I was just frightened. Mother was the boss — *always*.

12. Other significant adults. Tell me about any other adults who
    were important to you when you were a child. Describe the
    character and role of other adults living in the household in
    your pre-adolescent years. Also describe other adults who
    impressed you as a child in your pre-adolescent years, either
    positively or negatively (e.g., neighbors, teachers, family
    friends, etc.).

My father's brother, Uncle John, was a very funny man. I
enjoyed him. He always had something good to say about me, and
treated me as if I had important ideas and interests. He had a big
house and a big garden. He was a plumber and made a good living.
Also, I'd go with Dad to see my great aunt when she was in the
hospital. She was a very sharp, touchy woman, and difficult to talk
to. She'd given me a violin, and when we went to see her she'd
always ask me about it. We kept up the lie that I was doing fine
with it, though in fact I didn't play. It was Douglas who played,
and I had given him the violin.

Then there was Aunt Jill. She was a big, no-nonsense woman,
not affectionate, but not unkind. She just wouldn't hesitate to tell
you what to do.

On my mother's side there was Aunt Carla. She's the second
oldest of mother's brothers and sisters. She was a large, woman, a
mother-earth type. She ran a small business, a newspaper and
magazine stand. She always sent presents — magazines and books.
That was kind of nice. She never married. And there was my
mother's brother, Ned. He was called the "black sheep," but I never
knew why — still don't. I thought he was very nice and generous.
He never married either.

13. Family milieu. If not already clear, inquire into the other
    particularities of the childhood situation: socio-economic,
    ethnic/racial, religious and cultural characteristics of the
    family, and the family's standards and values.

We felt we were better off than the others in the neighborhood. Our house and yard looked nicer, we ate better, and I was better dressed. Of course, nobody had a car or TV — it was during the war, and it was a poor neighborhood. Even in our situation some of the rooms in the house were shut off sometimes. It was a fairly rough area, and Mother didn't want me to play with the other kids. She couldn't keep me from it, but she didn't like it and she'd lecture me about it.

Ethnically, I suppose it was about as mixed as it gets in a working class neighborhood in a city in Virginia. Of course, there weren't any black people. Not in those days! Dad was a kind of "Archie Bunker" type in his thinking. You know, "Jews are all moneymakers, out for themselves." All the standard prejudices. The funny thing about it was that, of course, he didn't really *know* any Jews, except for the people who ran the dry cleaner's, and with them he was always the soul of good manners. So I grew up with all that, and I've had a lot to learn.

I was steeped in the Episcopal Church. Sang in the choir and so forth, although it dawned on me as I got older that we were a little out of our league there, in terms of class. But it was part of Mother's thing. She always referred to us as "Anglicans"! My father didn't go, by the way — another bone of contention between them, and another opportunity for her to lord it over him. But my father was a religious man, in his own way.

He was into simple pleasures. He read all the racing sheets and put a few bucks on the horses now and then. Mother hated that. She'd always say, "If he weren't a gambler we'd have more money." Being honest was very important to my mother, and gambling didn't fit with that. You should be hard-working. Don't look for charity, and don't look for luck. Take care of yourself, keep your problems to yourself, be suspicious of your neighbors, and believe in Heaven.

SIBLING ARRAY

DAN (45)		DOUGLAS (-3)	
- I don't know what kind of a kid I was.		Much more popular than I was. He was easygoing. . . just took things as they came.	
Very conscientious.		He was well liked.	
Very intent on being seen the "right" way: - good behavior - polite - not rough - a gentleman		Easy to play with. He did what I wanted him to do.	
I was very talkative. I tried hard at school.	(Note: Mother had "some" miscarriages between the births of the two boys, and possibly before Dan and after Douglas: "I'm not too clear on this.")	He never made any trouble.	
I was bullied a lot by neighborhood kids beginning about 8 or 9. They were rough. I'd get my head kicked. My mother would always run to the neighbors: "Look what your Johnnie did to my Dan!"		He was in excellent health.	
I was sick with asthma from age 2 until I was 16 or 17. I usually couldn't take part in gym. And I was allergic to chlorine, so I couldn't swim in the summers.			

Family Constellation, Part II: The Situation of the Child

Sibling Array

Note: See the facing page for an illustration of how the data for Dan and his brother was entered in the sibling array.

Directions for using the Sibling Array form: A blank form may be found on p. 19 of the *IPCW*. Array the names of the client and siblings across the page from left to right. Include deceased siblings and siblings separated from the family, living in institutions or elsewhere. Also record Mother's other known pregnancies terminated by miscarriage, medical abortion, or still-birth, entering these in their appropriate ordinal positions. Using the client's age as the baseline, note the difference in years, plus or minus, between the client and each of these siblings/events. Include step-siblings on the form, recording the client's age at the time step-siblings entered the family, and age differences between step-siblings and the client. Use the items below to guide your discussion with the client, placing appropriate information about each child in the boxes provided. If there are more siblings than can be accommodated, draw a line across the form to create additional boxes.

14. What kind of children were you? Describe each of the children, beginning with yourself, with respect to personality, health, and activities.

15. Note any sub-groupings among the children: Who played with whom; who fought and argued with whom; who looked after and took care of whom; who taught and guided whom.

   We played together a lot. We scrapped (I'd win). The issue was who'd be boss. Douglas did what I said.

16. Record how each of the children distinguished himself or herself from the others, for example, by taking the place of the one who was:

academic/Dan   the problem child/neither
athletic/Douglas   religious/Dan
artistic/Douglas — violin;   socially successful/Dan
  Dan — painting ("I'd   developmentally disabled/
  paint with Mother.")   neither
entrepreneurial/neither   the good child/both
                               sickly Dan

17. If any of the children was developmentally disabled or sickly,
    note the child's and family's attitudes toward the difficulty.

I had asthma and allergies. I was hospitalized once when I was
four. We coped. I wouldn't get more than a few days off from
school. It was "tough it out." There was a "carry on" attitude.
About my teeth, Mother's solution was, "Don't smile!"

18. Who among the siblings were you most like? Least like? In
    what ways?

Like Douglas: We were both musical, though he really got into
it, and worked to learn the violin. We both had tempers, and still
do. It was the only way to get any relief from Mother's nagging.

She'd back off, but for me the price was high. I hated her cold
silences. We also both took schooling seriously enough to finish
college.

Unlike Douglas: He was more easygoing, more laid back than I
was. He gave Mother the impression that he was going along with
what she wanted, even when he wasn't. He'd just say, "Yes,
Ma'am," and then do as he pleased. It amazed me. All kids said
"Ma'am" and "Sir" to adults in those days — at least, in the South.
He'd fidget when he was nervous. I'd blabber. I'm a workaholic,
and I think he's got his life more in balance.

19. To what extent did each of the children accept, reject, or
    modify the family's standards and values?

We both went along, for the most part. Douglas was a little less controllable, especially when he got older, but generally we both went along. We always put up a front of being good.

20. What were your favorite stories, TV shows, movies, fictitious or historical characters? What was there about them that you liked?

I was a reader. I liked adventure stories. I read all the Hardy Boy books. I remember the kids. I read the Nancy Drew books, too. My aunt who had the newspaper and magazine stand got them for me. I liked Nancy Drew. She was different, spunky, ready for anything, daring. She'd get into scrapes. She was gutsy! She went where the boys went.

21. Did you daydream? What about?

I didn't daydream, but I had terrible nightmares.

22. Do you remember any night dreams? Describe. How did you feel when you woke up from the dream?

I don't remember what they were about, but I had them, and I'd be terrified. I remember Mother's telling me about a nightmare she had when she was pregnant. I don't know if it was when she was pregnant with me or when. There was this huge spider coming up over the end of her bed. It would keep coming, and she'd wake up in a sweat.

23. Did you have any particular fears as a child? How did others respond?

I was very afraid of spiders. I wonder if it had anything to do with that nightmare? I was afraid they'd get on me and devour me. Mother and Dad were nice about it, sympathetic.

24. Was food and eating an issue in the household? In what way?

You had to eat everything on your plate. And you had to stay at the table until you finished it. We just did it, that's all. That's the way it was.

25. What were you good at? What did you most enjoy?

I liked acting and drama, painting, drawing, and singing. Later I played softball. When I was a kid I roller skated. That was when I was asthmatic, and couldn't do a lot.

26. Was anything particularly hard for you? Was there anything you did not enjoy?

Lots of things. School. Arithmetic. Writing. It was never quite good enough. I talked too much, and I had some trouble with being accepted and making friends.

27. Did you experience difficulty in your mental development? Your emotional development? How was this addressed in the family? In school? How did you feel about it?

Well, that's a pretty broad question. I suppose if I did have some difficulty, I wouldn't have known it then. We wouldn't have noticed! It was "tough it out" all the way.

28. Think back to your early years, up to age six or seven. What did you want to be when you grew up? What was there about this that appealed to you? Did this idea change? How and when?

I didn't have a clue. Once we had to write a paper called, "A Day at Work." I was about eight or nine, I guess. I wrote about being a shoe salesman! My attitude was, "Just do the job." Oh, another thing: I loved trains. There were lots of trains when I was little. I loved the engines. Big, huge, steam engines. You'd feel the earth shake when they went by. I used to write down all the numbers of all the engines. I considered being an engineer, but it wouldn't have been considered the right job by my mother — or anyone. It was manual labor, and I saw myself in an office. Not that I knew what people did in offices!

29. Describe how you got along in the world of the neighborhood. How would you characterize your role among the children (e.g., leader, follower, jester, outsider, etc.)? Did you have friends of the other sex, of a different sexual orientation, or of a different ethnic/racial identity?

I was O.K. in the neighborhood when I was little, just one of the kids. When I got to be eight or nine I started getting bullied. I was identified as a "sissy" then, and never shook the label. I was always shy and ill-at-ease around the others, especially the girls. We didn't play with girls on our street.

30. Describe how you got along in the world of the school. Was your role in the school different from your neighborhood role? If so, in what ways? How do you account for it? Did you have friends of the other sex, of a different sexual orientation, or of a different ethnic/racial identity? How did you get along with teachers? What were your favorite and least favorite subjects, and what was there about them that you liked or disliked?

Well, in school we were separated when we got to junior high. They called it "tracked," and it was supposed to be very progressive. I was in the top group — in the middle of the top group. The kids from my neighborhood weren't in it, so it threw me in with another type of kid. Actually, they were the rich kids from other parts of town. So my group from then on was very different. It was hard, and it made my life in the neighborhood harder than it had been. After my teeth got broken they started calling me "toothless Danny." That hurt.

One thing that affected me all the way through school, especially in high school, was my accent. Real southern, or "southren" I should say, with an emphasis on "backwoods southren." When I was thirteen I tried out for a part in a play. They said, "Your acting is great, but your accent is terrible." I sounded "common," so I didn't get the part. I was stunned.

About teachers: I got along with them all right, except that I was too talkative. I jabbered. It was nervousness. My favorite subject was science. It was so orderly. And, I could memorize everything and learn it easily and get good grades. And I liked the projects. I didn't do too well in arithmetic; later, though, I liked algebra and higher mathematics. I liked sorting out the word problems.

31. Describe your bodily development in childhood, and how you felt about it. How did it compare with that of your peers? Consider height, weight, strength, speed, coordination, vision, hearing, any anomalies. Did you have any special difficulties (e.g., bed wetting)?

I bit my nails all of my childhood, right to the quick. I was average, neither too short nor too tall, fast enough. Worst for me were my teeth and my curly hair. I got teased about that. I wanted straight hair. I had a tendency to be skinny, but that was O.K.

32. Describe your sexual development and your sexual experience and initiation. How did you learn about sex? How did you feel about the bodily changes that took place at puberty? (Females: How old were you at the menarche? Describe what happened. Did you understand what was happening? How did you feel about it?) Describe your experience as a sexual person during adolescence and young adulthood, and your evaluation of yourself at that time.

I didn't have any sex education. It was unheard of then. All us kids in the neighborhood talked to each other. If we asked our parents, they'd ignore us or tell us the dumbest things. My mother actually said that I came out of the doctor's "little black bag." I didn't learn about intercourse until I was thirteen. It's hard to believe we were so ignorant! But when I did learn about it, it seemed reasonable to me. We had cats. It made sense.

What didn't make sense was to think about my parents: "They've seen each other without any clothes on! They've touched each other!" I was intrigued. I thought, "Gee, my parents have actually had physical contact."

When I hit puberty I ran into trouble with girls. There had been a couple of girls in grammar school I'd liked. When I was about fourteen, I began some sexual exploration. We'd feel each other under the desks — put our hands under each other's clothes. That was as far as it went for me.

I didn't actually start dating until I was about eighteen, and I didn't sleep with anyone until I was twenty-five, when I was engaged — later it fell through. Anyway, we had intercourse. I think it was probably awful for her. I was so reticent. It was so mechanical. I just kept thinking, "Quick! Something's going to go wrong!" I just rushed through it. I was terrified that I'd get her pregnant. That fear was part of what always kept me away from women.

33. Childhood chronology. If not already clear, note any changes that took place in the course of the client's childhood and adolescent development and situation that were experienced by the client as major events or turning points, whether for good or ill, and the meanings the client assigned to each experience.

The main thing was that we were both different from the other kids. We wanted to go to college, to get out. That made us truly different.

The Case of Dan (continued)

Summary of the Family Constellation

[Note: After reviewing the above data (Items 1-33) point-by-point to clarify an understanding of Dan's training and self-training for his approach to current difficulties, the consulting therapist dictated the following summary to the recording therapist in the presence of the client. This procedure is further described in the Introduction to Part IV.]

Dan grew up as the older of two boys, psychologically the firstborn son, in a family in which the atmosphere was one of disturbed security. The security lay in knowing that Father would provide for and protect the family; the disturbance lay in knowing that Mother would, nevertheless, be anxious, unhappy, and dissatisfied.

The family values emphasized by Mother and Father included hard work in caretaking of what one has (house, garden), and religious feeling.

The masculine guiding line set by Father stressed a kind of relaxed cheerfulness and steadiness, together with a temperament that was generally balanced, and a practical intelligence capable of flashes of brilliance. Father was also capable of putting up a show of fighting back, and knew how to pursue his own interests quietly even when opposed (playing the horses). Grandfather confirmed the masculine guiding line in his loyalty to the women in his life. Uncle John enhanced the masculine guiding line through his sense of humor and relative material success.

The feminine guiding line set by Mother stressed a fretful, ambitious, pessimistic, and unpredictable busyness. Mother was better than others, and yearned to be among her betters. Mother's failure to express warmth was confirmed as characteristic of femininity by the attitudes of Dan's great aunt and his Aunt Jill, both of whom were cold. Even Aunt Carla, a generous, comfortable woman, kept her distance by "sending" presents, and by not making an alliance with a man.

Dan stayed ahead of Douglas by sharing Father's interests, and by being Mother's chief object of interest. Mother, who saw her life in tragic and heroic terms, was determined to see to it that her sons would succeed in ways that had been foreclosed to her by what she regarded as unfair circumstances.

It was difficult for Dan to understand Mother's fretfulness about appearances, about striving, about education, about moving up and out in the world, and about Father and his alleged inferiorities.

In the world of the home Dan apparently imitated Father's way of quietly enduring Mother's agitation, which led to his enduring her fussing over him as well. He also imitated Father's temper outbursts, by means of which he (and Father) were able to gain occasional relief from Mother's pressure. In Father he found a source of acceptance, but no standards by which to measure progress. Mother, who praised him when he met her standards and chastised and withdrew from him when he did not, provided the source of judgment.

Douglas, three years younger, made a place for himself by leaving the first-place position to Dan, uncontested; by staying out of trouble; and by developing an easygoing, friendly attitude toward life. Seeing that Mother's focus on Dan was the source of Dan's troubles, Douglas contrived to stay out of Mother's way, and out of her "spotlight."

Dan found another source of acceptance in his Uncle John, whom he often saw in Father's company. John encouraged Dan, took pleasure in the boy's efforts, recognized his intelligence, and, like Father, did so without judgment.

In the world of the neighborhood he was able to do well enough without striving, buttressed by a sense of being "a cut above" the others that carried him through his difficulties with them. His good appearances, maintained by Mother, and the good appearance and steadiness of the household, probably sustained him in this idea.

He managed well enough in the world of the school until adolescence, when everything changed: He now found himself propelled into a new group of peers, and compelled to assess his personal situation against the background of a much wider social world.

Dan suddenly saw himself as a very small duck in a very large pond. His uncertainty about his place in the larger world, beginning at this time, continued to express itself in his hesitation to engage with the other sex in caring intimacy, until in his mid-twenties, when he could make his first effort to do so under the protection of a formal engagement.

In adolescence, all of Mother's complaints against Father's lack of ambition, and all of Mother's warnings against Dan's being "like the Allens," previously hard for him to understand, came to make a certain kind of sense to him: Appearances in his new world were much more important than he had thought. It was not enough to look good; he would have to sound good as well, with the proper accent. He came to see that Mother's brand of ambition and striving were essential if he wanted to better himself. (Unlike Father, he could not afford to miss any buses; for him, another might not come along.)

The question that confronted Dan at the threshold of adult manhood remained: How could he prove himself to be a man unlike the Allen men? To do this, he might suppose, would require that he set out in quest of an unhappy woman, and then demonstrate that he could make her happy by becoming whatever she wanted.

Record of the Early Recollections

Before we return to the lifestyle assessment inquiry in "The Case of Dan," it will be useful to offer some observations on the process of eliciting and recording early recollections. (See Chapter 11 for a further discussion of early recollections.)

An early recollection may be a complete story with a beginning, middle, and end, or it may be an account of a single moment; it may describe activity, or it may be limited to the memory of a sensory perception.

For the purpose of a lifestyle assessment, an early recollection is defined as a memory of a single incident of early childhood experience, preferably referring to occurrences in the first nine or ten years of a person's life, prior to the onset of puberty.

Note that this definition distinguishes an early recollection from any report of an experience that does not focus on a single occurrence.

A phrase such as "I remember one time . . ." or, "Once when I was . . ." indicates the introduction to an early recollection.

Reports, by contrast, characteristically begin with "We used to . . ." or, "We always . . ." or, "Lots of times I . . .", phrases that do not refer to a focus on a particular incident.

Reports often contain early recollections which, with a little prompting, may come into view. If you ask for an early recollection and the client proffers a *report*, simply inquire, "Are you remembering something that happened one of those times?" or, "Do you remember one time when that happened?" If the answer is no, proceed to an early recollection by asking, "What is the next thing you remember from your childhood?"

Make a marginal note of any reports which you receive in the course of recording early recollections. They may stimulate further recollections during the review and interpretations, and they can help to clarify a context for the interpretation of the early recollections themselves.

Gathering the Record

Begin the process of gathering the record of early recollections by telling the client: "The next thing for us to do together is collect a few stories from your childhood, from before the time you reached puberty, up to, say, age nine or ten. I'd like you to tell me some incidents you remember from that period, beginning with the very first thing you can remember."

The lifestyle assessment inquiry anticipates that a record of eight recollections will provide an optimal amount of material for interpretation. It is of course possible, and sometimes necessary, to work with fewer early recollections.

It is most important to strive, as far as possible, for a verbatim record of the early recollections. If a client likes to give long, discursive accounts, this may prove to be difficult; with a little practice you will learn to distinguish the recollection itself from irrelevant introductions and stage-settings. The goal is to obtain first-person narratives using the client's exact language. Example:

> *Early Recollection:* I was about three, playing in the front yard. As far as I can remember, I was alone. There was a fence around the yard, a picket fence. There I was, making mud-pies.
>
> Most vivid moment: My hands full of mud, sort of squeezing it between my fingers.
>
> Feeling: Just great! Fabulous!

After the first recollection is recorded ask, "What's the next incident that comes to mind?" Make it clear to the client that the memories need not be recalled in chronological order. Accept whatever arises after the recollection just presented. There is an internally coherent, artistic arrangement in the sequence of the early recollections which is, in itself, of psychological importance, and therefore of value to a psychological investigation. This sequence should not be disturbed by the imposition of a temporal ordering. (See Chapter 14 for a discussion of this artistic arrangement.)

When the client has presented eight recollections say, "We have all we need for this record. Is there any other memory that is important to you, and that you haven't had a chance to tell so far in our work together?" Often a client will say something like, "Well, there is just one more," and go on with, "I thought of this a little while ago, but it didn't seem important enough to mention"; or, "Yeah. I didn't want to tell you this one, because it's sort of unpleasant [weird, embarrassing, scary, etc.], but I think about it a lot. Now that you ask, I guess I might as well tell you about it."

There is a tension involved in holding a memory back and wondering whether to report it. When you say that you have enough memories in the record, the tension is relieved, but the client remains uncertain whether it makes sense to withhold information that might better be explored. So, when you present this further opportunity to get it into the record, the tension will often break, and the client will confide material that enables you to get a fuller understanding.

Special Problems in Gathering Early Recollections

1. *Not remembering*

Sometimes, even before we begin the inquiries in a lifestyle assessment, the client will report being unable to remember anything from childhood. The review of family material usually stirs reminiscence; by the time we arrive at the recording of the early recollections a complete store of incidents is almost always available to the client's memory.

Sometimes, however, a client will not remember specific incidents from before the age of eight or nine, or even later. Do not make an issue of it. Stay with the format of asking for the earliest memory, and then recording the others as they are presented.

Bear in mind that memory serves to maintain the individual's *orientation* in life. In most cases the basis of this orientation was established by four or five years of age, and the first memory dates from this period. Failure to remember back before a much later age can be taken as the sign of a *disorientation* that occurred, and of a *reorientation* that began at that time. For example:

> A client could not remember an incident from before he was twelve years old. He knew about his life before that age, and could report that he had grown up on an isolated farm. As the much youngest of three boys he was his mother's companion and helper in the house, while his brothers worked the farm with his father. He attended a one-room school where he was one of fourteen children of various ages. When he was twelve he began to go by bus to a consolidated junior high school, where hundreds of young people were enrolled. For the first time in his life he found himself in a group of boys who were age mates, among whom he was able to take his place in team sports and other activities.

This man's failure to recall specific incidents from before the age of twelve reveals the disorientation and reorientation that he experienced then. The impressions he had of self, others, and the world before he was twelve were no longer useful. It was necessary for him to evaluate everything afresh in the new situation, which he had done successfully. He provided a rich set of early recollections from the period between ages twelve and sixteen. (Later, as his work in therapy progressed, he began to recall incidents from his earlier years, and was able to integrate their meaning into his new awareness of himself.)

2. *Absence of affect*

When asked, "How did you feel at that moment?" (the "most vivid moment" in the story), a client will sometimes say, "I don't know." A feeling expresses an *evaluation*. If I tell you how I feel about something, I tell you where I stand relative to that issue. Those who have rehearsed the posture of not letting others know where they stand are often unable to recall feelings in their early recollections. Do not press for a feeling when none is reported, and, above all, do not suggest to the client what he or she "must" have (or even "probably") felt in the situation. Empathy is required of you, as is the courage to guess; the aggrandizement of another person's subjectivity is forbidden.

There are other motives for being unable to recall feelings:

> A man of forty-two recounted eight early recollections with no hesitation. He was, however, unable to recall a single feeling in connection with any of these incidents, even though the recording therapist had patiently encouraged him to take his time to focus on each memory. During the review of the material with the consulting therapist, he readily identified feelings for each of them. The recording therapist was a woman, the consulting therapist a man. According to this client's masculine and feminine guiding lines, the judgment of a woman was not to be trusted under any circumstances, and he would not entrust his judgments to a female therapist.

3. *Inclusion of family constellation early recollections in the formal record of early recollections*

During the task of recording early recollections, clients will refer to incidents that were recorded during earlier inquiries, or that came up and were discussed during sessions of review and interpretations. A client may say, "I'm thinking about the time my grandfather and I were picking apples. I already told you about that. Do you want me to go on to another story?"

We say, "No. Let's include that here." Then we ask the client to tell the story again while we enter it into the record at that point.

4. *Historical accuracy*

Sometimes people say, "It probably couldn't have happened this way, but this is the way I remember it." Assure the client that you are interested in the way he or she remembers experiencing the incident, and that historical accuracy is not important for this purpose.

The client may say, "I'll have to talk to my sister about this. I'm not sure about it." You can respond that this would be interesting, but that for the purposes of the assessment, it isn't important to know how anyone other than the client remembers the incident. (It will be useful for such a client to discover that others remember incidents differently, since those incidents had a different significance for them.)

5.   *Dreams as early recollections*

The client says, "I remember this dream I had when I was a kid. Should I tell it to you?" The answer is yes; dreams are experiences. Any dream qualifies as an early recollection if it is remembered as taking place during the childhood years, and can be recalled as a single incident. This is true also of recurring dreams which the client had over extended periods of time, even many years, provided that a single occurrence can be recalled.

6.   *Photographs as early recollections*

The client may wonder, "I don't know if I remember this, or whether I only know it from a photograph." Here, there are two possibilities: The client has reconstructed a memory of the incident during which the picture was taken, or the client is recalling an incident of seeing the photograph. Ask, "Can you picture yourself in that situation in your mind's eye? What's happening?" The client may respond,

> Yeah. I'm standing in front of the swing — like in the photo. I'd just gotten off, and my sister's been pushing me. She's not in the picture. She's standing by the tree. She's angry because I won't push her, and I'm kind of pleased with myself for having gotten away with it.

Other times the response is,

> "I can't get it. All I see is the photo." If the client recalls the incident of seeing the photograph (of looking at it, of discovering it, or of being shown it by someone), and if this took place in childhood, the early recollection will be something like this: "I remember the first time I saw that picture. I was sitting on the sofa with Mother, going through an album. She pointed it out to me. I was pretty excited."

7. More than one "most vivid moment"

When asked, after recounting an early recollection, "In all of this, what is the most vivid moment?" the client replies, "I remember two things vividly. The first is riding the bike by myself — the moment when I was aware Uncle Bill had let go of the seat, and I was on my own. The second really vivid thing is when I lost my balance and fell over and the bike fell on me." Ask how the client felt at each of the two moments, making a record of both responses in the numbered space for the early recollection.

8. "Bum rap" and "good rep" stories

One may recall being told a story about oneself in the course of one's childhood and being unable to recall the incident which the story claims to portray. Such a story may burden the client with a "bum rap" (that is, a charge of faultiness based on allegations of which the client feels innocent, or at least unaware). Another such story confers a "good rep(utation)" on the client (which can also be experienced as a burden, something which the client feels obliged to live up to).

> A woman recounted a family story that is told about the time she dumped her baby sister out of her carriage, causing the baby to cut her head, and requiring that the baby be taken to the hospital for stitches. Although the client could not recall this incident, she had felt bad about it all her life. Whenever the sister failed to perform well at anything, some family

member could be counted on to refer to the accident: "It's your fault. If she hadn't hurt her head this wouldn't have happened." Even though there was usually a light-hearted, joking quality to these statements, the client felt anxious and uncomfortable in the face of them. It was as if she were being reminded that she was, inescapably, a person expected to cause harm, an idea she tried to compensate against by being overly solicitous and cautious.

A man reported that, although he cannot remember the incident, it appears that he saved his brother from certain severe injury and possible death by grabbing the younger child as he was about to fall through an open upper-story window in the family's apartment. "I dislike hearing that story. I'm no hero, and I'm not a guy who thinks and acts quickly. Every time I hear that story I feel like they're talking about someone else. I guess it happened, since they all say so, and my father was there at the time and saw it, but I always feel inadequate when I hear it. I'm an ordinary guy, and they expect too much from me."

These "stories about me," which are not subjectively remembered as part of one's own experience, but which can be imaginatively pictured, may be thought of as "imposed early recollections" or "imposed personal myths." The effect they have on a person's self-image and sense of what is required in life may invite strategies of safeguarding, or other special struggles to meet the expectations of others. While they are not authentic early recollections, we must take them into account when we are framing the early recollections summary, *The Pattern of Basic Convictions.*

Directions for Recording the Early Recollections

Following is the sequence we use for recording ERs: (1) ask the client for an early childhood recollection, (2) record the account verbatim to the extent possible, (3) narrow the range of the ER to the moment most vividly recalled, as if in a snapshot, then (4) ask how the client felt at that moment.

Prompting questions: How far back can you remember? What is the first incident or moment you remember in your life before age nine or ten? Tell me about it. How old were you at the time? What is the most vivid moment in the course of the story? How did you feel in that moment?

Note that the first ER to be recorded is to be the first incident of memory. ERs recorded after the first one follow the logic of the client's memory, so may not be in chronological order. After recording the first ER, ask, "What is the next thing you recall from before age nine or ten?" or, "What comes up next from childhood before age nine or ten?" If the client cannot recall any ERs (or as many as eight) from before age nine or ten, accept and record whatever ERs may be presented.

The Case of Dan (continued)

The Record of the Early Recollections

[Note: The material below follows the form of the *IPCW*, pp. 26-29.]

ER 1. Age 3 ½ _____

I remember when my brother was born. It was cold and dark, January. The weather was bad. I remember the doctor coming in — it was common for babies to be born at home in those days. He had a big black bag. I remember being alone outside the bedroom door and not being allowed in.

> Most vivid moment: Standing outside the door; the door
>   was ajar.
> Feeling: Frustration. I thought, "There's something going
>   on in there that I'm not allowed to
>   share in or see. Why aren't I?"

ER 2. Age __3__

It was during the war. I was playing in a sandbox in the yard.
There was an air-raid drill. Do you remember those? It was
afternoon, the sky was blue. Mother came out and rushed me into
the house. I didn't want to go in.

> Most vivid moment: Mother coming out looking very
>   worried, picking me up and taking
>   me in.

> Feeling: Frustration. I thought, "Why don't you let me do
>   what I want to do, which is to stay here?"

ER 3. Age __4__

Sleeping over at a friend's house in a basement. Sort of a
recreation room, but it hadn't been finished. It was cold and damp
down there.

> Most vivid moment: Going down into something dark,
>   damp, unpleasant.

> Feeling: Reluctance. I didn't want to go.

ER 4. Age __5__

I'm in the kitchen with my mother. There's a spider on the
ceiling. My mother sees it and she freezes.

> Most vivid moment: Mother being immobilized.

> Feeling: Real insecure. I was aware that my mother was
>   scared, and that made me feel insecure.

ER 5. Age <u>5 or 6</u>

We were all out together, Mother, Dad, me, and Douglas. We came to a busy intersection. Dad got Mother to hold my hand and Douglas's hand. We stepped out to cross the street. All of a sudden a car swerved toward us. Dad sort of tackled us all and pushed us out of the way, shouting, "Look out!" He almost got hit himself.

> Most vivid moment: Dad being capable. He was calm in a way: He just did what he had to do. He just acted. Threw himself at us to make sure we'd be O.K.

> Feeling: Safe. I thought, "My dad's strong and protective and capable."

ER 6. Age <u>6 or 7</u>

I was walking to school. I noticed that a building that had been across the street from my school wasn't there any more. There were bulldozers there in this hole pushing dirt around, and the building was gone. It had just vanished!

> Most vivid moment: Seeing that enormous hole and realizing that the building was gone.

> Feeling: Wow! I felt a sense of wonder at it. A tinge of excitement. Sort of, "Look what people can do! Look what Man can do!"

ER 7. Age <u>8</u>

At school we had a fire drill. Our teacher, Miss Johnson, led us down a long corridor and the line slowed down. I was standing in the hall, waiting for the line to move. I saw an empty socket, and I stuck my finger in it and got a shock.

Most vivid moment: Getting shocked.

Feeling: Embarrassed.   I pretended that nothing had
             happened and didn't tell anyone.   I thought,
             "They'll think I'm stupid."   My arm tingled and
             hurt all day.

ER 8. Age 5

My brother fell into the lake at the park. He was a toddler, two
years old. We were playing by the lake together, sort of squatting
down and playing on the concrete edge by the side of the lake. I
looked up and saw this coat floating on the water. I just stood there
and looked at it. I was aware that Douglas wasn't there, and that his
coat was in the water, and I remember thinking, "But where is he?"
Then someone grabbed the coat, and there was Douglas.

Most vivid moment: Seeing the coat floating, with Douglas
                     nowhere in sight.

Feeling: Astonishment! All I saw was the coat — it didn't
           register with me that Douglas might be under it, so
           I was astonished. He was gone, and there was his
           coat.

After gathering the eight ERs, inform the client that a sufficient
number has been recorded. Ask if there is any other recollection
the client would like to have included in the record; if so, make a
record of it using the same format, designating it "ER 9." Use the
same format, featuring "Age," "Most vivid moment," and
"Feeling."

[Dan had no additional ERs.]

Early Recollections Summaries

Summaries of the lifestyle assessment include the summary of *The Pattern of Basic Convictions* and the summary of *The Mistaken Ideas*. At the conclusion of the review and interpretations of the ERs, the consulting therapist dictates these summaries to the recording therapist in the presence of the client. See Part IV of the text for a discussion of the way early recollections are reviewed and interpreted, and the way summaries are developed. Also see pp. 51 – 55 of the *IPCW*.

The Case of Dan (continued)

The Pattern of Basic Convictions

Life is centered around the mystery and wonder of Woman.

I don't know how I fit into it. Somehow I feel kept from entering into acquaintance with the feminine heart of things, as if I were not allowed to share in and to know the mysteries of intimacy even when the door is partly opened to me.

At the same time, I feel unable to move freely on my own in the world, under the open sky, because of the power a woman is able to exert over me.

I don't want my life to molder in some narrow place for the sake of my friendly feelings toward someone.

A woman is able to throw fright into me by her being frightened. There is no need for it. Being afraid is no help against danger anyway.

A man knows enough to take sensible precautions, and to do what he can and must do in the face of difficulty.

I am impatient with being shepherded by a woman. I want to find out for myself what life is all about. If I have to take my lumps I am prepared to suffer in silence, especially when I am embarrassed by my difficulties. I'm afraid that anyone to whom I might confide my pain would fail to sympathize and may only think that I'm stupid.

Even so, I am in awe of what I am discovering about the wonders of human power and human possibility.

There is more to life than meets the eye, at least my eye. I am astonished at how much I have failed to see, when life itself is at stake, and it is imperative that I learn to see clearly. I am aware that I have needed the assistance of others to save me from disaster, and I can take some comfort from the thought that nothing goes so far wrong (as long as we're still alive) that it can't be set right again.

The Case of Dan (continued)

The Mistaken Ideas

[Note: After allowing time for the client's reaction to *The Pattern of Basic Convictions*, the consulting therapist immediately dictates an enumerated statement of the client's *Mistaken Ideas*, in the client's presence. See Figure 7, p. 406; also see p. 56 of the *IPCW*.]

1. He has some idea that the door to intimacy and belonging is closed to him when, in fact, he knows it's open.

2. The meaning of closeness with a woman is mixed up in his mind with the idea of his freedom being restricted and his courage being undermined by her fears.

3. He has some notion that his sense of wonder and curiosity is an idiosyncrasy, and that he cannot expect to find sympathy or understanding for the difficulties he experiences because of it.

4. He has an exaggerated idea about masculine competence, as if he believed that a Real Man can do everything, and he is left feeling like a wide-eyed boy in a world of wonderful men.

Conclusion

In this chapter we have set forth the process of gathering family data and early recollections according to the format of the *IPCW*, illustrated by "The Case of Dan," and have presented summaries of the material. We return to Dan's case in Chapter 13 (pp. 381-387), where Dan's family constellation summary serves as the model on which we base our discussion of how this section of the lifestyle summaries is developed and presented.

This brings us to the end of Part II, "Inquiry." The next step in the psychoclarity process is the review and interpretations of the material with the client. Part III, Chapters 5 through 11, is devoted to discussing and explicating the various aspects of this task.

PART III: REVIEW AND INTERPRETATIONS

Anyone who wants to understand Individual Psychology correctly must orient himself by its clarification of the unitary purposefulness of [the] thinking, feeling, willing, and acting of the unique individual. He then will recognize how the stand an individual takes and the lifestyle, which is like an artistic creation, are the same in all situations of life, unalterable until the end — unless the individual recognizes what is erroneous, incorrect, or abnormal with regard to cooperation, and attempts to correct it. This becomes possible only when he has comprehended his errors conceptually and subjected them to the critique of practical reason, the common sense — in other words, through convincing discussion.

Alfred Adler (1979, p. 52)

Individual Psychology and the psychoclarity process based on its findings do not allow for a sharp distinction between diagnosis and treatment. These terms fit a model of interaction in which one person seeks a thorough knowledge (diagnosis) of another's state of being, and then applies an appropriate corrective (treatment) to the other person's disorders.

The "convincing discussion" that Adler refers to on the preceding page is, by contrast, an inherently collaborative effort. The corrective to one's own psychic suffering and sense of social maladjustment is gained by a "thorough knowledge" which recognizes and understands the errors in one's own private sense. Paradoxically, however, this most personal of responsibilities cannot be exercised on one's own. A collaborator is needed, skilled in the task of subjecting hidden errors "to the critique of practical reason, the commonsense." The errors in the client's private sense of the world are fatefully and hopelessly pitted against the community; the therapist serves, therefore, not only as a faithful worker with the client in the struggle for clarity, but also as a representative of the community, negotiating an end to hostilities.

The meaning of this encounter can be misunderstood. It is not between a therapist who is all commonsense, and a client whose sense is all private. No discussion of any kind could occur if the client did not share in any way in the sense we have of our common life, and of what it requires of us. No convincing discussion will occur if the therapist does not know, through firsthand experience, what it is like to be guided by biases, and blind to one's own errors — the psychoclarity process does not take place between a therapist above and a client below.

A little quatrain by Rudyard Kipling (Bartlett, 1968, p. 187) warns us against the uselessness of such a posture:

The toad beneath the harrow knows
Exactly where each tooth point goes;
The butterfly upon the road
Preaches contentment to that toad.

In simplest terms, the collaboration of client and therapist is in itself the fundamental and primary corrective to errors in the client's private sense, which are "erroneous . . . with regard to cooperation." The effort to understand the lifestyle is already contrary to the maintenance of a private sense, since understanding is by definition something that can be shared. Here, diagnosis is treatment.

This is why everyone who writes on psychotherapy and counseling of any kind emphasizes the importance of *rapport*. To us that word means the alignment of goals, or working together in quest of the same objective. This is not always easy to maintain. The anti-cooperative errors in the private sense of a discouraged person distort the meaning of any movement toward collaboration. You can expect to be set up as an expert and put down as a failure, all in one deft move.

An advantage, therefore, of the psychoclarity process for assessing the lifestyle is that you and the client are engaged, right from the start, in the pursuit of something the client wants. In the review and interpretations you are drawn together into the work. The client, wanting to be understood, and having to help you to understand, begins, almost inadvertently, to share an understanding with you.

A caution is in order here. Remember that people seeking any kind of psychological counseling are likely to be confused and disoriented. Disorientation is hard to endure. It leads to disorganization and a frantic, wasteful dissipation of energies. This is why survival trainers warn us that if we are lost in a wilderness, the first thing we must do is sit down, focus attention, and take our bearings. Panic is a threat to life. In the pain of it, people are suggestible. When we lack clarity, we crave certainty. Fanaticism (both political and religious), esoteric cults, and arcane pseudoscience of all kinds appeal to this craving. Charlatans feed on it. Psychotherapists can exploit it.

For example: A client was told by a clinician that he was "a depressive." He took this in as information about himself, but not without a response. He evaluated the "data" and gave it a meaning. To him, this was not an academic matter. Even a depressive has to live and make his way, and he had to figure out how. He formed an expectation ("I will be sad"), and an ambition ("I will bear up under it nobly to demonstrate my goodness, apologizing for its inescapableness as an excuse for the trouble it causes others"). To those who sought to cheer him up, he offered a game try at a cheerful face; if his inconsolability was felt by them as a reproach, he pleaded the excuse of his "disorder." It was the best he could do. As his therapist had said, he was, after all, "a depressive."

It is better to avoid labeling. Stick to the task of understanding the person's *training, self-training, and rehearsal* for life as it has taken place until now. When an understanding is gained, and the social meaning and consequences of the movement are clear, the symptom can make sense to the client; it is only by means of such understanding that anyone is able to consider changes.

The client comes to the therapist with "problems." As Powers has put it, part of the therapist's task is to reveal these problems to be *solutions*, faulty solutions, to be sure, to the problems that confront us all: making friends, making a living, and making love (Griffith, 1984, p. 439). Each of these problems requires of us that we cooperate with others for their solution. The client's "erroneous, incorrect, or abnormal" convictions, upon which the faulty solutions rest, have left cooperation out of account. The substitutes for cooperation all involve some form of coercion. With respect to each of the three problems of life set for everyone, it is as if the client were privately thinking something like the following:

Making friends: In order to make people like me, I have to be (beautiful, good, intelligent, impressive, rich, strong, etc.) enough to be irresistible to them.

My problem is that I am too (plain, bad, dumb, ordinary, poor, weak, etc.) to be sure this will happen.

Making a living: In order to get as much money as I want I have to have (a boss who appreciates me, a helping hand, the right connections, a better education, the ability to do everything right, the proper knack or talent, etc.).

My problem is that (my boss is a jerk, nobody gives me a break, I don't know the right people, my parents wouldn't pay my tuition, I get nervous and make mistakes, my brother inherited my father's abilities, etc.).

Making love: To warm up to someone I have to know that the other person is (considerate of me, kind, sensitive to my needs, undemanding, reliable, respectful of my freedom, etc.).

My problem is that I keep running into people who (are out for themselves, selfish, want things their way, are unwilling to meet reasonable expectations, are not there when I need them, refuse to bend, etc.).

Lacking here is friendliness, eagerness to make contributions, and an interest in enhancing the lives of others and sharing affection with them as they are. What is left is a pattern of manipulative strategies, claims to positions (or conditions) deserving of special treatment, self-elevation, excuses, suspicions, and reproaches against others for their imperfections.

Until now, the client worked things out tolerably, and the faulty ideas of a hidden private sense were never challenged effectively. The ideas did not invite clear challenges, because they were never expressed in open statements, only in a general movement.

Whenever the effects of that movement were opposed by others as onerous, they could be defended by the declaration of good intentions, by rationalizations, and by appeals for greater sympathy (as in Edward's case, discussed in Chapter 2).

Now the individual confronts or anticipates a situation that will not yield to the usual strategy of movement; the style brings this person to the narrow end of the wedge. Clients know something is wrong; they do not know what it is, or what to do about it. They are addressing a current difficulty in the same way in which they have dealt with all their previous difficulties in life. This time there is no resolution. In this situation they are lost. We have all heard of the man whose only tool is a hammer; he tends to treat everything as if it were a nail. If the situation fails to respond to his hammering, his only idea is to hammer all the harder.

Lifestyle operates something like that. We have trained ourselves to respond to life's problems in a particular way. If a problem will not respond, and we do not know another way to solve it, we hammer all the harder in our practiced way. We can only do what we know how to do.

During the review of material gathered in the inquiry, therapist and client work together to learn how the client came to his overreliance on hammering, and to consider how other skills, undervalued until now, are available to the client to be studied and practiced. The chapter that follows gives an example.

Chapter 5. THE LIFE SITUATION, PRESENTING PROBLEM, AND LIFE TASKS

This chapter consists of a transcript of a review of initial interview data, together with the text of the *Summary of Impressions* derived from that interview. Except for minor changes made to protect confidentiality, it is a verbatim transcript.

As the conversation proceeds, you will be able to see the development of the ideas proposed, tested, and modified as we go along.

This follows Adler's (1963) description of his work:

> I may very well, at first, make an error in interpretation which will come to light later on as the case unfolds. If so, it will not discourage me. I am aware that I am in the same position as a painter or sculptor, who at the outset does whatever is suggested to him by his experiences and skill. Only later on does he check his work, strengthening, softening, and changing the features to bring out the correct image. (p. 1)

The Case of Harry

[Note: After the recording therapist gathered the client's data, using the initial interview format of the *IPCW*, and after a brief break, the client ("Harry") and both therapists, Robert L. Powers (RLP) and Jane Griffith (JG), convened to discuss and clarify the information.]

JG: Now, Harry, what I'm going to do is tell Bob what I've learned about you. If you want to say something more, or if I didn't get it right, please interrupt me at any point.

Harry: O.K.

JG: And you and Bob will go off into your own conversations as we go along.

RLP: For example, if I don't understand something, I'll ask you to clarify it for me. And, it'll be interesting to you as we go along to hear whether Jane did "get it right."

Harry: (Laughter.) O.K.

JG: Harry is thirty years old. He was born July 20, 19xx.

RLP: So you had your thirtieth birthday two months ago. How do you feel about being thirty?

Harry: It's O.K., but sort of unsettling.

RLP: In what way?

Harry: I don't know. Well, probably because everything is unsettled right now.

RLP: Maybe we'll want to come back to this topic of your birthday. For now, let's go ahead.

Harry: O.K.

JG: Harry's been in Chicago for four months. He moved here from Davenport.

RLP: That's just across the Mississippi from Moline, isn't it?

Harry: Yeah. That's my hometown.

JG: He's currently in partnership in a printing business.

RLP: That's a business foreign to me — except as a consumer.

Harry: It's foreign to me, too. I've had a lot of professional dealings with printers before, but I've never been on the inside before. I'm learning the business strictly from scratch. I came here without a mentor, though there are people who are helping me.

JG: You'll learn more about this bold act as the story unfolds. (JG and Harry laugh.) He's not married, and has never been married. He's having an on-again, off-again, in, out, hot, cold love affair with a woman named Debbie who is getting her M.B.A. degree at the Wharton School. They met in college at Ohio University. He's known Debbie for a long time. However, it's only been a little over a year since they actually began their love relationship. Before that, she was dating someone else who was in the group Harry was in.

RLP: So you were social acquaintances for some time, and became lovers about a year ago.

Harry: Right. It started when we saw each other at a party. She was looking for somebody to take a trip to Florida with. I said, "Fine, I'll go," knowing full well that maybe this would be a chance to start a relationship with her.

RLP: Oh.

JG: Harry completed his degree at Ohio in economics, then returned to Davenport to join the family business. He's not been in the military. When I asked about his religious affiliation or interests, he said, "I'm Jewish. I'm not very religious, but I am very Jewish. I have strong ties with the religion." He says that he's not a practicing Jew right now.

RLP: But you clearly identify yourself as a Jew.

Harry: Well, I'm very involved with Jewish affairs. I guess I'm a practicing Jew. I'm just a practicing Reform Jew . . .

RLP: Oh, I see.

Harry: . . . as compared with most of my relatives, who are very Orthodox.

JG: You don't practice to that extent.

Harry: Yeah. Not to that extent. I guess I am pretty religious.

RLP: So you're observant — as much as you can be in the culture.

Harry: Right. Right.

JG: He says, "I'm very intense over the fate of the Jewish people and Israel, though I'm slacking off on that." He then told me about the family business in Davenport: "I was in a good family business, making a lot of money, and I was getting involved in Jewish community affairs. However, I didn't get along with my father on the job, so now I'm trying to do something else that appealed to me." He was in the family business for eight years. "I was miserable, and felt I should get out. Debbie was the catalyst. She gave me the push, because she said that she would never want to live in Davenport." Since his move, she says to him, "Don't try to put it off on me that you left." He says that he doesn't do that, but she sometimes feels like she is responsible, because of the ultimatum she gave him.

RLP: Uh-huh.

Harry: All along she said, "Don't make your decision based on me. I don't want you to do that."

RLP: Because it wasn't contractual. She wasn't saying, "If you leave, then I'll marry you."

Harry: Right. On the other hand, she was saying, "If we do stay together, it can't be in Davenport." I mean it's a "Catch-22."

RLP: It's possible that we'll stay together, but not in Davenport.

Harry: Right.

RLP: But, if you leave Davenport, it doesn't add to the possibility of our staying together. That's not, by itself, sufficient cause.

Harry: Right.

JG: This is the main issue about which Harry is here today: What's going on between him and Debbie. He says, "I want to marry her, but she seems to be throwing up roadblocks. But then, the funny thing is that when I back off, she's all over me." For example, if he doesn't call her for a couple of days, she calls him and "has eight or nine things she wants to tell me."

RLP: Where did you say she is?

JG: In Philadelphia. At the Wharton School.

RLP: Is she the same age as you?

Harry: Yeah. She's a few months younger.

JG: His problem is that he can't concentrate on his work, make his reputation, make his new career work out for him, because he is too distracted by the situation with Debbie.

Harry: Did I say that? Did I say reputation? I don't care about that. (Laughter, all join in.)

JG: No, that's my word. I meant, to establish yourself as a success in the new business.

RLP: Making connections, and so forth.

Harry: Right. Right.

JG: Debbie keeps saying, "I can't tell you what I am or am not going to do." She blew up at him just the other day because over the weekend Harry got together with some of their group of mutual friends, and he was surrounded by happily married people. It made him feel like he wanted to move on the whole thing. He kind of challenged her on the phone by telling her, "I've given more than you have, and whenever I do something that you want me to do, then you throw a roadblock in my way." And she blew up.

Harry: We've been having a problem where she felt like she's fallen out of love and isn't attracted to me the way she was, but now I'm her "best friend" type of thing. I wanted to confront her, and I said, "Well, why have you fallen out of love?" And she says, "I can't tell you. I really don't know why. I don't know what's happened. Maybe it'll get better." She'd be very happy to talk about other things, but not that. And I finally kept pressing and said, "I'm having a lot of trouble up here, and when you've had trouble I've helped you through (and this, that, and the other), and you haven't been a big help to me right now. And besides that, every step of the way you've thrown up another roadblock. First it's, 'Well, I can't see you this weekend'; that's a roadblock. Or, now you're telling me you're not in love with me, but you don't know why." And I finally got her to say something, and she said, "You always expected more out of this relationship, more than I have," and she left it at that. She really didn't talk about it any more. But that pretty much is the situation I'm in. I have pushed from the start. Like I said I saw that trip as a chance to start a relationship, and . . .

RLP: So you had high hopes and ambitions of its developing . . .

Harry: Yeah. I still do, and I still push. And, she likes me to push, but then the pressure gets too much. She likes certain aspects about it. So that's what happened: I pushed too much.

RLP: She likes being courted, but doesn't want to be wedded.

Harry: She doesn't like being tied down. Really, at this point she doesn't have any desire to go out with anybody else — or it doesn't seem that way. But she doesn't like the idea of her freedom being squashed. I don't know.

RLP: We mustn't spend too much time figuring Debbie out. You're the one who comes to talk with us; she doesn't.

Harry: (Laughter.) Yeah. That's what I said to her. I'm spending all my time trying to figure out what she's thinking, and what I need to be spending my time on is coming up with something productive for myself to deal with, especially at work.

RLP: Yes, sure, because your life is at stake.

Harry: Yeah.

JG: Mother and Father are both living. Mother's now fifty-five, and Father is sixty. It's the original marriage for both of them. They've been married thirty-two years, and live in Davenport. Both are in good health. Mother has a job working in a lawyer's office. His father is in a grocery business that Harry's great-grandfather founded. There are some other relatives who also partly own the business.

Harry: Yeah. His brother owns some of it.

RLP: The grocery business?

Harry: Right. Wholesale.

JG: Harry is the first of four children. He is followed by Sidney, his brother, two years younger. Sidney is in the family business now and is having some of the same experiences with Father, but is responding to them differently — as we could expect. Gail, a sister, is about three years younger than Harry, and Burton is seven years younger. I asked him about the groupings of the children. He said, "We were one group. We were all treated pretty well." Gail wasn't special, even though she was the only girl. Harry and Sidney "got along pretty well," though they fought a lot. But that's "just growing up." When I asked how much younger Gail was, he wasn't really clear about it. He said, "Well, I guess about three or so years younger. Something like that."

RLP: The first three of you were born pretty close together, weren't you?

Harry: Yeah.

JG: Burt was a lot younger, but Harry still thinks of this as one group of kids. He doesn't think of Burt as a tag-along, younger child. About his own position as the firstborn, he said, "I paved the way."

RLP: That was your job!

Harry: (Laughter.)

RLP: No wonder you have the daring to go off on new ventures, to launch out into a new world. You're used to it. You were raised to pave the way.

Harry: Well, we've all pretty well done that. My brother Sidney, for instance . . . I've more stayed at home in some ways. I feel like I ran back home! When I finished school they said, "You've got to go to work." So I said, "I'll go to work here." I stayed there. My brother, he left school a number of times, and he went and traveled. Sold whatever he had and took the money and traveled. And he finally started in the business when he was in his late twenties, and he's able to cope with it better.

RLP: You paved the way.

Harry: Yeah. Yeah. I guess . . . yeah. For somebody to go and do that stuff.

JG: I asked Harry what he would have wanted different, if he could have had something different when he was a child, and he said that it was a pretty good life for him as a little boy growing up, that things got more difficult later.

We then talked about his health. He said, "I'm not as energetic as I'd like to be. I'm too heavy." I asked when he perceived himself to be overweight, and he said, "I lost a lot of weight two years ago. About a year ago, I went to Florida with Debbie. Then, some months later, I started putting it back on. I was bummed out with my situation and I started to binge." He sees himself still doing that. "I still don't exercise. I was agonizing over leaving my job and making the move." Generally, he sleeps well. Recently, in the past two months, if he doesn't sleep well, it's because he's thinking about his job.

RLP: Two months ago you turned thirty. Things should be settled by now, shouldn't they? By the time your father reached thirty he was married and had a child.

Harry: Right. Yeah. That's true.

JG: We talked about alcohol, drugs, and so forth, and he doesn't perceive himself as having any difficulty with these, though he said that, among things of this kind, cocaine could be the biggest problem. "I never did it a lot, and since I've been in Chicago, it hasn't been a problem. But, you know, if I'm with people who are doing it, I'll probably do it." He doesn't, though, see himself as someone who would go off into a cocaine oblivion.

RLP: A lot of people do.

Harry: Yeah. I know. But I think it's changed. I don't see the guys around . . .

RLP: They're not around. They're gone!

Harry: (Laughter.) It does bother me, because I think I could become an addictive-type personality. If somebody threw it in my face, or gave me the money for it, maybe I'd do it. But at this point it's certainly not a problem for me. I guess I could blow my life savings on it, and in six months I'd be over and done with.

RLP: It's a very seductive drug, because it presents an enlargement of possibilities. It brings about a drug-induced mania that's very attractive to people.

Harry: My bingeing on eating has kept me away from it.

RLP: Bingeing on eating is a little different. It's more an effort to console yourself for a feeling of emptiness. You recognize it?

Harry: Yeah. It makes sense.

RLP: Because you do feel empty. You have an empty heart. You want someone to embrace. You can't fill your life, so you fill your stomach.

Harry: And what's the difference with a cocaine binge?

RLP: That's to satisfy the ambition to be elevated . . . to be great and to be grand.

Harry: I could satisfy that by doing my job right.

RLP: Sure you could, but that takes work! Cocaine is a shortcut to those feelings. People with high ambitions are vulnerable to the cocaine habit. It gives them a feeling of power, grandeur, majesty.

Harry: Yeah. That's the feeling. But I've never had the desire to do it at work, or take massive quantities of it like I've read about.

RLP: It's dangerous. A dangerous drug.

Harry: That really is interesting . . . the difference between cocaine and food binges.

JG: Not only is there the Debbie-related emptiness, but here's a person who's uprooted from his whole world he's known all of his life.

RLP: Family, community.

JG: Yes.

Harry: I was wanting to get away. I was even bingeing there. I felt empty there, too.

RLP: It wasn't satisfying to you.

Harry: Especially near the end, when I was going to the psychiatrist.

JG: I'll be telling you about that. First, I'd like to finish this about his current health. You should know he's not under a doctor's care. About a year and a half ago his blood pressure was a little elevated, and required some medication. Then it got O.K. He's not taking medication now. He feels that he should have this checked out. He has a physical every year, and will be having one soon. About the psychiatrist, I asked him what he had found out. He said, "I found out I was stubborn — which I already knew — and that I don't take criticism very well — which I also already knew. It was a disaster."

RLP: How much did you pay to find these things out?

Harry: (Laughter.)

RLP: I always wonder when they say things like that, who does take criticism well?

Harry: (Laughter.) Well, I did find out some things about myself. The reason it was a disaster was more from the fact that it didn't help the situation. My father always harped on that business about my not taking criticism very well.

RLP: Right! That means without having you complain!

Harry: That's what it's always been.

RLP: Yes. He wants to feel free to criticize you without your complaining about it.

Harry: Yeah. Those were the things my dad was saying . . . that I wouldn't take criticism. The psychiatrist was trying to go back and get to the deeper roots of the relationship, what's really happening. He finally said that I was reaching out for my father's love, and I wasn't getting it from him. But that was totally over my father's head.

RLP: Over your father's head . . .

Harry: Yes.

RLP: He couldn't understand why you would want it.

Harry: (Laughter.)

RLP: Well, he was innocent. He didn't know what was going on either. What you want is for him to respect you. You want him to treat you as an equal. You don't want always to be in the position of the one who is being criticized by somebody who knows better. You also want to have something to offer.

Harry: Yes. And we had a business relationship and he couldn't draw the line between business and personal. And he made no attempt to. The main thing I was supposed to remember was that I was the employee, and that he was having to run the business.

He'd always say, "Well, this can't be run this way or that way." But you (JG) probably have this down.

JG: As a result of the counseling — which lasted about three months — Harry said, "The doctor and I understood what the problems were, whereas my father wouldn't see the problem. He would just dismiss it. It aggravated me and made me mad. The outcome was, the doctor said, 'Well, you can't get along with your father, and you ought to leave. You are making the effort, and your father is not responding.'"

Harry: Not that I was making the effort, but that I understood the problem better. But even if both of us had understood it, it may not have become any better. But, with him not understanding it, it really wasn't getting any better, because I was still just going back and, even understanding it, I was still doing the same things. It made me even more upset that he didn't understand it, that he just dismissed the advice.

JG: Harry's response to this idea that he ought to leave was, "I never wanted to leave the money. I still don't feel I've been dealt with in an equitable manner." In terms of the money situation and this feeling, Harry said, "In the company I was pretty much doing everything he was doing. He'd be out of town a lot, and I'd be running the company. My brother came to work then, and gave me an out. When I was there, whatever went wrong was my fault. Everything's got to be perfect. Father would rail on my 'mistakes' but his mistakes were 'judgment calls.'"

Harry: Now that I hear that, I can probably make a more accurate statement. His feeling was always that the bottom line was that no matter who made the mistake, it comes back to you how the business operates. If somebody makes a mistake, or something goes wrong, it's your fault eventually, because you're going to have to deal with it. In that respect he was right. He always felt there was a right way and a wrong way to handle a thing, and the right way was always his way.

RLP: Sure.

Harry: It was, "It's not your way or my way; it's the right way to do it." But the right way was never my way; it was always his way. He wanted that. And, my mistakes were always very objective mistakes. They were very clear-cut as to how I'd made a mistake. But when he made the same mistake, it was a "judgment call," and maybe it wouldn't even be a mistake. And a lot of my mistakes weren't mistakes in the end, either. There's no doubt that I can't run the business the way he does after thirty years. I suppose there was an intense rivalry between the two of us.

RLP: Well, some of it's on his part. He had some fear, if he felt it was so important that he keep such a close eye on everything.

Harry: It's everything I read about small business. They can't let go.

RLP: It's hard. He has some sense that it's up to him to maintain control of it. There's some pessimism in it, some fear that, "unless I hang on and watch everything, it'll all go to pieces."

Harry: That's exactly what he said: "Two years from now there wouldn't be anything here." And he feels that way. When he gets really angry and says what he feels, he says, "This place won't be here in two years. I'll go to Arizona."

JG: Father also said that Harry was lazy, and Harry said, "I was." He lays it to resistance to Father's constant criticism and nagging. He says, "From high school on they saw me as a problem child," although this is something Harry really doesn't understand. An example he gave me was, "I went with a girl older than I was, who'd been divorced. That drove them crazy. They wouldn't say anything; they'd just stand around being nervous."

RLP: How old were you?

Harry: Eighteen. She was twenty-two. It was a work situation. I can understand why they thought I was a "problem child," although I never thought that I was.

RLP: Not at eighteen. You're not a child any more.

Harry: (Laughter.)

JG: Then at college. They told him, "We were worried about you at college," when he wouldn't go to classes or something, and they'd find out about it.

Harry: In my third year, I stopped going to classes for a while, and they got a card from the college listing all my incompletes. They said, "It must be drugs."

JG: They told him, by way of explanation, "You're a member of that generation."

RLP and Harry: (Laughter.)

Harry: And then they said . . . my mother still believes this to this day! When I left town, she had a good cry. Her boy was leaving town. And I told her I was just so angry with my dad and all, and she said, "We've always known you were a member of that generation." This is after I'd been working in the business for eight years! Well, "that" generation is at least five years older than me, if she's talking about the sixties, and that's what she was talking about. She just . . . she's out of touch.

RLP: She's frightened.

Harry: O.K. If it's that simple.

RLP: Sure. Sometimes there's a kind of pessimism that anything that can go wrong will go wrong. You don't have to be Irish to believe in "Murphy's Law." Even a Jew can defer to it. Your parents have the time-honored idea that they can make life better by warning you against making it worse.

Harry: (Laughter.)

JG: Harry says about all this "old business" between himself and his parents, "I file all this stuff away, and never forget it." He's got a great litany of difficulties he's had with his mother and father.

RLP: Yes, because it feels like a "vote of no confidence" to him.

Harry: Right. Being away from them makes it a whole lot easier.

JG: He went on to say, "Now my only problem — since that's not so pressing — is the anxiety of a new job and the situation with Debbie. I'm not sure if I'm in control or not."

RLP: I don't understand what that means.

Harry: Well, I'm not sure either. (Laughter.) The problem is, I'm not dealing with it. It would be real easy to say, "Well, I've got to stop thinking about what Debbie is thinking about, and just go on and do my job better." But I stand around speculating about what Debbie's thinking all the time, and how am I going to make the situation right. And I go into my job and I don't do what I should be doing there because of that, and I'll turn around and go home and eat a lot, and I won't feel like going to work the next morning. Then I'll go out of town and be with my friends over the weekend and have a good time, and come back thinking about that. You know. . . it's just . . . that's my problem: I can't do what I should be doing.

RLP: No, you can't because you're wrapped up in the idea of control. It's reflected in your endless thinking about what Debbie is thinking, and your question, "Why have you fallen out of love with me?" This is, by definition, an impossible question to answer. Why anybody falls in love with anybody can't be very closely answered. It's about as clear a choice as there is, to fall in love with somebody, but it's hard for us to account for it. "Why?" is always a hard question to answer in any case, because it leads to speculation. It is especially to be avoided in intimate relationships, where "Why?" feels like an adversarial question. "What?" and "How?" make for an easier set of questions. They refer to facts, to things that are givens, things that invite a further decision. You and your father, though, aren't content with those two questions.

Harry: (Laughter.)

RLP: You and your father are both interested in trying to control events, and to manage the thoughts and feelings of those around you. In your case, if Debbie doesn't think what you want her to think, you're left to wonder, "How can I get her to think that way?"

Harry: That's right. That's exactly right!

RLP: You and your father operate the same way. You're a slice off the old salami.

Harry: I hate to be told that. I was told that once before. We were told that we're so much alike that we just react like each other.

RLP: Well, sure. If you're looking for apples, you look under apple trees.

Harry: (Laughter.)

RLP: You grew up under his tutelage. The two of you want to be in charge, and that's part of the conflict between you.

JG: So the control — or the lack of it — in the situation with Debbie is similar? That's the situation that precipitated Harry's calling us.

Harry: Yes, that and the fact that I'm not performing. I'm not giving my job a chance.

RLP: True, but Debbie is the distraction from it.

JG: Yes.

Harry: I think so. You know, I want to make sure that I don't find something else to distract me. I want to discipline myself more. I'm just thinking about it, instead.

JG: Thinking instead of acting.

Harry: Yes. That's what I'm doing. I'm not taking action.

RLP: No, you're thinking *instead* of acting. If you knew what to do, you'd act; since you don't know what to do, you think about it endlessly. That's what obsessive thinking is.

Harry: That's right. That's right.

JG: He said about Debbie, "I need a better way to deal with it. It's fifty-fifty whether it'll work out. I need to cope with it." Then he kind of pictured a scenario: "I'll be this loving boy who'll get used up and thrown out on the street. I'll let her take advantage of the situation. I have a fear that it won't work out. And what I should be doing is having a good time right now, and not be thinking about long-range commitments, marriage, and all that stuff. I was really hurt this weekend, being with my close friends who are happily married. Debbie rarely feels the way I feel, and she rarely wants the same things I want." Then he itemized a few things: "First, it was leave Davenport. She would get nervous living there all of her life. She was ecstatic about it when I left."

Harry: Yeah. She was.

JG: He went on, "Then, it's that she's not sure now. She says she doesn't love me, and she's not sure what happened to that feeling." And third, "She keeps throwing up road blocks." Then (and this is Harry, quoting himself to Debbie), "You've needed help. I've been there for you." And, she responds to him, "You've always wanted more out of this relationship than I have." He commented to me parenthetically, "Debbie is insecure. She had a hard time in school. I know she's competent, but everything always gets to her." She wanted an M.B.A. and applied to a number of places.

Harry: Yeah. And she was accepted at two places that were top-ranked. Now she's at Wharton, and she feels that she lied her way into it.

JG: She says, "I bull-shitted my way in, and now they're finding out that I'm nothing."

Harry: Here I am spending all my time worrying, getting into that whole rationalization and everything else. And I probably should just be letting it take its course.

RLP: I don't know what you should be doing, but . . .

Harry: I guess that's what I'm asking.

RLP: . . . but I know there are some other choices.

JG: I asked Harry how his life would be different if he didn't have this problem — this distraction with the unresolved Debbie situation — if she were committed to him, and he said, "I'd try to make a success of what I'm trying to do, then we'd pick a place where we wanted to go, and we'd get married. I'd be happier on less money, or maybe I could make the kind of money I want doing what I'm doing. Debbie doesn't want to marry a wreck."

RLP: Who does?

Harry: (Laughter.) Yeah. I don't blame her!

JG: He said, "She doesn't want to be part of my insecurities. I asked her to stand by me now, and she said, 'Fuck off!'"

Harry: NO! Did I say that?!

JG: Yes.

Harry: No . . . yes, *I* said it. Yeah. But *she's* not saying that. She's just saying, "Don't show me your insecurities all the time," which I can understand because she has so many herself. I think that bothers her.

RLP: She feels bullied by it.

Harry: O.K. O.K. That's probably it.

RLP: She's bullied.

Harry: By the insecurities?

RLP: Yes. You're bullying her.

Harry: I understand that. I really, fully understand that.

RLP: She doesn't like it. She feels pressured. More pressure, accusation, suffering. It all winds up to be bullying. Listen: "I stood by you when you needed me, but you don't stand by me when I need you."

Harry: Yeah, but that was a very short sentence.

(Harry, after a moment, laughs. Laughter all around.)

RLP: It doesn't matter what size it was. The point is that it did harm. She was hurt by it and offended by your bullying.

Harry: I know. I understand that.

RLP: It's like a very small bomb. Or a very small bullet . . .

Harry: (Laughter.) I understand . . . but I finally got her to say something that . . . that finally cleared up a little of my head . . . that made me realize that I was pushing her.

RLP: Yes.

Harry: She'd never say to me, "You're pushing me too hard."

RLP: You'd never say that to your father, either. But you felt it. Any pushing at all is too much. It doesn't work. In love, especially, it doesn't work.

JG: I asked Harry what he wanted to come out of today's meeting with us and he said, "I have to put Debbie in her own compartment, and sort out the work area of my life. I want to be able to not binge and not sit around obsessed with Debbie, or obsessed with anything that takes my mind off work, like bingeing."

After this we talked about love, work, and friendship from other points of view. I asked Harry about what kinds of qualities come to mind when he thinks about a "masculine" man, and the first picture in his mind was a bunch of body builders in a magazine advertisement. Then, "Well, I guess getting a lot of dates makes a man masculine."

Harry: I don't know. . . . Did I say that? It's hard for me to . . . it doesn't sound right.

RLP: O.K. What would you rather put in place of it? What does distinguish a man as masculine to you?

JG: Yes. You said, "Someone who is very sure of himself, someone who is aggressive."

Harry: Domineering . . . which is what Debbie called me. That's what she said.

RLP: "Aggressive" can become that.

JG: On the feminine side, "The way they dress — like makeup. I'm trying not to think of stereotypes. I want to deal with people on a personal level. But my ideas of masculine and feminine probably are stereotyped: probably the dominant person and the submissive person. But I don't like that picture."

RLP: Especially when it doesn't work. You've picked a woman who is not about to submit. Submission doesn't sound like Debbie's style.

Harry: No. That's true!

RLP: I don't suppose that M.B.A.s are submissive types, generally. They're decision makers, or at least they want to be. They want to take control of a situation — run empires. You said to me on the phone [a reference to his original request for an appointment] that she wanted to go into the investment banking field. We don't want our investment bankers to be submissive, do we? When we have a lot at risk, we want them to maintain tight control, and be decisive, I should think.

Harry: She's worried that she's not like that, but she is.

RLP: That she's worried that she's not, only shows her determination to be that way.

Harry: Oh?

RLP: Same for you. You're afraid you're not, which shows your ambition to be that way.

Harry: Hmmm.

RLP: Otherwise, you wouldn't think about being decisive or being in control at all. Not being decisive enough doesn't enter my thinking, unless being decisive is pretty important to me.

Harry: (Laughter.) Makes sense. I understand that.

RLP: That's the line of your ambition.

JG: I asked then about his first encounter with Debbie, and this is the story he tells: "I met her when I was in college, and she was dating a friend of mine. I felt, there's this beautiful girl, she's Jewish, I like her. We had similar backgrounds and interests. I wasn't love struck by any means. Then she was asking around at this party, looking for someone to drive to Florida with her, and I thought, 'Here's a chance to develop a relationship with her.'" This group of friends, he says, is quite clannish. Both of them have been in the group a long time.

He went on to say, "I was in some bad relationships prior to our getting together, and so was she. I saw something that I wanted, and I went out and got it. But now she's holding back. At this party — and I hadn't seen her in quite a while, and she had dated a bunch of other guys in between — we sat together and talked for a long time. We had a lot in common, and we had fun. I thought, 'I'd like to have a relationship with her — how can I do that?' Then this opportunity came up."

I asked Harry about his ability to express affection toward her, and receive affection from her: "I have no trouble expressing affection with Debbie. Probably I'm too aggressive."

RLP: There's a hint in there that you were sensitive to the fact that she'd also been in some bad relationships — perhaps with your friend — and you had some feeling that she shouldn't have to have that kind of trouble . . . such a beautiful, sensitive woman; that you appreciated her, and that she might want to have someone who appreciated her and enjoyed her. You said something that hinted at that.

Harry: Well, yeah. Why wouldn't she want that? I want that. I suppose everybody wants that.

RLP: That's probably true. But in her case you saw someone who had not been appreciated, and who might be especially grateful for your appreciation.

Harry: No, no, no. I didn't know she had had bad relationships until I knew her better.

RLP: Oh, you didn't?

Harry: I knew she was going out with some guys. I mean, I found out later, afterwards. Maybe that's tinged my thinking since.

RLP: So at the time you only knew that you liked her a lot.

Harry: Yeah.

RLP: That she's a very attractive woman.

Harry: Yeah. But it's not that she's such an attractive woman; it's that I'm attracted to her.

RLP: What is there about her that makes her attractive to you?

Harry: I really don't know what it is. Although lately, when I've been losing my grip on the whole situation, not letting it play out its course, forcing issues, I don't know what I like about any of that anymore.

RLP: Well, you're sort of angry, and . . .

Harry: Yeah. I guess I have to let it play its course, or whatever. I don't know what you were going to say.

RLP: Well, I don't know what I was going to say, either. I only know that in matters of love a good motto is, "If at first you don't succeed, quit, quit at once."

Harry: (Laughter.) Well — so, if she does turn around, what would happen? Would we have a miserable life?

RLP: Well, you would if this is a sample: on-again, off-again, back and forth.

Harry: That's what it is. That's what it boils down to. . . . "Quit, quit at once." (Laughter.)

JG: Harry went on to say . . .

Harry: "Quit, quit at once." That's great. O.K.! (All laugh.)

JG: He goes on to say, about receiving affection from her, "I'd love affection from her, but it's not there right now. It always was more me than her." He said that there was a time when they'd "rush off into the bedroom together," but now he says, "I guess that probably was more me than her, because she's just not sexual right now. She's turned me into a 'best friend.' And she's so nebulous about it. She says, 'I don't know what's gone wrong.'"

RLP: She probably doesn't.

JG: He added, "When I approach her physically, she says she needs more space; when I'm playful she says, 'Don't be a nerd.' She wants certainty in her life, and I can't provide her with that."

Harry: That's just me, rationalizing.

JG: He says, "I expect too much out of the relationship."

RLP: You want her in your life, and she does not want to be in somebody else's life. She's having trouble enough figuring out her own life. She was happy to have *you* in *her* life — she invited you to come on her trip! It didn't occur to her that you were going to ask her to come into your life — to come on your trip. The moment that idea came up, she said, "I don't want to live in Davenport," which was the symbol of it.

Harry: Uh-huh.

RLP: For you to be in her life was all right. For her to be in your life wasn't very comfortable to her.

Harry: That's interesting.

RLP: That's how it sounds.

JG: She wants you on her terms.

Harry: Well, yeah. That's what I've been trying to say: She sets the terms.

RLP: Sure. She thought you agreed to the terms. She wanted a companion for her trip to Florida, and you were happy to join her. It was attractive to you to have the chance to be with her, because you were hoping it would develop into something. That would have represented a new set of "terms." But it doesn't sound as if Debbie is at a point in her life where she's able to make an alliance with anyone. She's too busy trying to figure out how to complete her studies, her education, her training, and to establish her competence in her own field.

Harry: Well, this is a better way to rationalize the whole thing to myself. It makes sense to me.

RLP: If I am correct, you don't have to rationalize anything. You just have to recognize that she doesn't want to get married. She wants to be your best friend. Best friends want their best friends to find someone else to marry.

Harry: She wants to get me out of her life.

RLP: Yes, if you don't want to be friends with her, and all you do is bring pressure into her life in an effort to change the course of her life.

Harry: And she wants to be nice about it.

RLP: Yes, and, if you want to get married, she'd like you to find someone who wants to be married to you as you are, right now.

JG: Someone who will "take the pressure off me," as Debbie put it to you.

RLP: Yes.

Harry: It's not going to be enough just to ease up on her — and still be with her.

RLP: I don't think so. You've led her to believe that you want her to marry you, and your hope that she may come around to that is experienced by her as a pressure. You've led me to believe that she doesn't want to get married. She doesn't feel sure enough of herself, of her own competence in her chosen field. She has her own project to complete. She wants a sense of having completed something for herself.

Harry: Are you saying that from the way I've explained it, it's a doomed project, even two or three years from now?

RLP: I don't know. Two or three years from now, if you were to meet again, and if neither of you has found somebody else in the interim . . .

Harry: You're saying I shouldn't wait.

RLP: You can, if you want to, but I don't think you're going to join a monastery for two years, and come back out when the time is right. You may, but I don't think you will. In any case, this isn't the point.

   For anyone who stays in a love affair that's not working, we have to ask, "What's the purpose of staying in it?" There's a number of possibilities that occur to me: First, you're in the same circle, you and Debbie. There'd be some loss of face. You've gone after her, but she won't have you.

Harry: People are saying, "When are you going to get married?"

RLP: Right. "When are you going to get married," and so on. So there's some sense of defeat, of not being able to succeed, of having made a real effort and not succeeding.

Harry: Yes, and I also don't want to . . . it's going to be tough to throw myself back into the game, the . . . I don't know.

JG: The market?

Harry: Yeah. The market, or whatever you want to call it.

RLP: Well, the line is forty blocks long for a young Jewish man who wears a clean shirt.

Harry: (Laughter.)

RLP: And who brushes his teeth every day.

Harry: (Laughter.)

RLP: A line forty blocks long of young women, young Jewish women, who will be delighted to meet you and be happy with you *just as you are*.

Harry: Well, so then it comes back to, "but she's so perfect for me, and I'm so perfect for her."

RLP: Then we have to ask ourselves, "In what sense is she so perfect for you?"

JG: What kind of perfection is this?

Harry: (Laughter.) This grief. (Laughter.) What kind of perfection is this grief?!

RLP: We have to ask ourselves whether you want to be married at all.

Harry: If this is the kind of life I want, maybe I should stay single. (Laughter.)

RLP: At the moment, I think you do want to stay single. You are also at a time in your life when you want to establish your independence in your own field, in your own business. If I were in your place, anyway, I would want to be able to demonstrate to my father that I'm able to make it on my own.

Harry: Oh, sure.

RLP: And I think it would be such an important ambition for me that I would allow myself to be intimidated by it from time to time. If I want so much to show my father that I can make a success on my own, and if that's what I most want, what is it that I don't want? I sure as hell don't want to let my father see that I've made a botch of it on my own. Right?

Harry: Uh-huh.

RLP: That's the negative pole of the ambition, and it's enough to intimidate anybody.

Harry: I'm not so much worried about everything else.

RLP: Right. And if this is such a big deal, and if I *do* make a botch of it, it may be to my advantage to feel — in my heart, anyway, and to save my self-esteem — that I made a botch of it because I was caught up in an unhappy love affair that distracted me from it.

Harry: Here I've got the excuse.

RLP: Yes. I've got the perfect explanation for having botched it up.

Harry: That's what I'm saying. I've got a perfect excuse for botching up.

RLP: Sure. How could I possibly concentrate on my work? And there's more. There's a positive side to it as well, because if I do make a success of it, in spite of the fact that I've got an unhappy love affair, the success is a triumph — a success in spades!

Harry: Uh-huh.

RLP: So then I've proved that I can do it even when . . .

Harry: . . . the chips are down.

RLP: Right. Even when I'm handicapped. So the distraction with Debbie and the preoccupation with food and so on, is a kind of insurance policy. If I succeed, I'm a tremendous success, and if I fail, who could have succeeded?

Harry: So I've used it, in other words, like a tool.

RLP: More like an insurance policy. With this much trouble, plus a success, it's a triumph; with this much trouble and no success, it's not a failure. It's understandable. Who could possibly have succeeded?

Harry: Uh-huh.

RLP: Either I have a triumph, or I'm exonerated.

Harry: Uh-huh.

RLP: It's a marvelous policy. It pays in either direction. An insurance man once said to me, "There are two things you have to insure against: one is that you'll die too soon; the other is that you'll live too long."

Harry: (Laughter.)

RLP: And you have a similar kind of insurance policy: Either I'll succeed big, or I won't fail at all.

Harry: So I shouldn't . . . I mean, my life will go on the same whether I worry about it at all.

RLP: Well, I don't know what you should or shouldn't do. It's only important to see what, so far, you have been doing so that you're free to consider what else you may want to do. I think it's important for you to consider disengaging from this ambition to prove something to your father, at least to some extent.

Harry: Disengage from it?

RLP: From the ambition to show something to your father, to prove something to him.

Harry: Should I not be intimidated by it? You said, "Don't be intimidated by it."

RLP: I only pointed out that you are, and that I don't think it's necessary. I don't think it is worth it to make your life into a project for proving a point to your father. It may be more satisfying for you to find something you enjoy doing, and enjoy your own experience of competence.

Harry: I thought that was what I was trying to do.

RLP: I think the other ambition intrudes on it.

Harry: Oh, sure it does, a little . . . or some. A lot, maybe.

RLP: Enough to spoil it for you.

Harry: It spoils the job right now.

RLP: The fun.

Harry: Yeah. The fun. Yeah! I've always enjoyed managing, selling — until I got here. It's certainly not as easy as I thought it was going to be either.

(Long, thoughtful silence.)

JG: Would you like me to go on? O.K. [JG continues.]

After talking about love, we talked about friendship. Harry has friends he grew up with and went to college with who are here. His best friend is Bill, who lives in San Diego. He also has good friends back home, "but if I told them too much it would get back to my parents." Also, they bug him with the question, "When are you going to get married?"

Harry: It's all in relationship to her. My parents and everybody are always asking, "How's she doing?" And that's the . . . hey! That's the defeat!

RLP: Uh-huh. You can define it as a defeat, and then you'd be terribly distressed. Or, you can define it as a choice: "I don't choose to pursue her any longer. She has her own agenda, and she has to work it out on her own."

Harry: Uh-huh.

JG: I asked Harry about what kind of impression he thinks he makes on people when they first meet him and he said, "Well, I talk. I'm talkative." That's about as far as he could go. He said, "I don't have a clear picture of that. I do have an image of myself, though. I'm real insecure. I act much more sure of myself than I feel and sometimes I feel that this, that, and the other thing is wrong with me."

RLP: Isn't this in line with the masculine ideal of decisiveness?

Harry: Uh-huh.

RLP: See, "I act much more sure of myself than I feel," because men, Real Men, are supposed to be clear, decisive, and untroubled about their decisions.

Harry: Uh-huh.

RLP: Whereas, as a mortal person, I'm not always so sure. I make choices without being sure. But there's some idea that I'm supposed to — as a man — have confidence in my decisions and know what I'm doing. That sounds like Father talking.

Harry: Yeah, sure. (Laughter.)

RLP: There's a right way and a wrong way, and you should know the right way, and so on.

JG: When he said, "Sometimes I feel that this, that, and the other are wrong with me," he said, "That could be a little paranoid, but when I look back at my history, maybe there's some evidence that there is something wrong with me."

RLP: That's a haunting idea. I have something to say about that. Is that the end of your record?

JG: No.

RLP: Well, let me just say this briefly. There's something wrong with everyone. Every human body has anomalies of structure. And no two human beings have the same organic strengths and weaknesses. When we make an issue of it, we turn our ordinary imperfections into a syndrome.

I want to write a paper on this subject some day. The difficulty is in trying to capture a tone of voice and a mood in the title of the paper, which I plan to call, "The *Something Wrong* Syndrome." This requires an upper case "S" and an upper case "W" and the words have to be italicized, to suggest the haunting quality of introspection and self-doubt involved in the syndrome.

In fact, there is something wrong with everyone, but we usually refer to it in "lower case" terms, and without the italics. Most of the time we don't focus on our deficiencies, or on what they may prevent us from achieving. Instead, we focus on what we *can* do with what we've got, and count on the contributions of others to assist us when necessary, or even to correct us when that is required. The *Something Wrong* Syndrome stalls us, and keeps us busy in a search for imperfections, as if we had to be faultless before we could proceed with anything. That's not possible.

Harry: Yeah. Well, you're saying it's another excuse.

RLP: Something like that. But not a very good one.

Harry: (Laughter.)

JG: Back to the record, all right? I asked Harry what his friends value in him. I asked, "If I called Bill and asked him what there is about you that he values, what would he say?" Harry said, "He'd say I'm a good friend. We do a lot for each other. When a person needs something, Bill or I, we help each other." Harry added that he thinks of himself as a loyal friend. He doesn't run out on his friends when the going gets rough or something goes wrong.

In the work area, I asked if there's anything he brings to his work that he's aware of as troublesome. He said, "Yeah. If only I wouldn't run away from it. If only I'd stand there and do it. I get side-tracked." Then he said, "If there's something that I really want to do, I do it."

RLP: Like the rest of us.

Harry: (Laughter.)

JG: He said, "I felt Dad really threw a lot on me, and I feel bitter about that. Finally, I would just say to Dad, 'You do it!' I don't stick to things. There are things that I let go." Then he added, "There was just too much for one person to do. Dad had brought the business out of a swamp — the place was going broke, and he rebuilt it. He could have bought out the rest of the family and he didn't. Now he has nothing to leave his kids except his share of the business, and he's bitter about this. I would scream at him when I was in the company and we weren't getting along. My brother has a different attitude. He says to Dad, 'O.K., we'll do it your way,' and just smiles."

RLP: A secondborn.

JG: Harry also said, "My dad feels terrible about what's happened. A number of times he came home crying." When he and Father got along, though, they went sailing together. Harry's father is a great sailor, and they have a big sloop. But Father felt that Harry should learn to sail it. Harry said, "He felt he had to push me to take the boat out. He nagged me, and I resisted and I kept putting off doing it." This, even though Harry wanted to do it, and felt able to do it, and wanted to take his father out.

RLP: You took each other too seriously, the two of you. You're too impressed with each other!

Harry: (Laughter.)

RLP: He's very impressed with you. Therefore, he's frantic to tell you everything he knows, and to see that you learn everything. With your brother he's more relaxed. He's not so impressed with him — and Brother isn't so impressed with Father, either. He says to Father, "Fine. O.K. We'll do it your way." Then he goes and does what he wants anyhow. That's my guess.

Harry: Well, he does what Dad tells him to — with the job. In his personal life, though, he doesn't do anything Dad tells him to.

RLP: No. It doesn't matter to him.

Harry: That's right. He doesn't listen. He just walks out, leaves. He says, "Screw it!"

RLP: Sure, but you and your father are too impressed with each other. Each of you is trying to impress the other, in turn, that you also know something. Father is so impressed by you, he's afraid that you won't be impressed by him.

Harry: Uh-huh.

RLP: You're so impressed by Father, you're afraid that Father won't be impressed by you. Do you recognize what I'm saying?

Harry: Yeah. Yes. I see what you're saying. It's a form of rivalry.

RLP: No. It's not a form of rivalry. It's a form of affection. You really admire your father.

Harry: Oh, yeah!

RLP: For what he did for that company, and what he did for his whole family . . . taking that company out of the swamp that his father had left it in. You admire his loyalty to the other members of the family. Am I right or wrong?

Harry: Well, that was stupidity for the most part. He really feels that it was.

RLP: That may be, but it's not what *you* feel.

Harry: No. But, yeah. I admire his morality and his motives. I have a lot of respect for the way he thinks.

RLP: The fact that they abused him is not to *his* discredit.

Harry: Oh, I see what you're saying. Sure I do, I admire him.
RLP: You admire him. And what do you want in turn? You want him also to admire you. It's the same for him. He admires you. You're his firstborn son. He sees wonderful things in you. You can do anything. When he sees hesitation on your part, he says, "Come on! You can do it!"

Harry: (Laughter.) Yeah. You're probably right. Yeah.

RLP: Yes. It is right. Otherwise, he wouldn't be so frantic. Part of your difficulty is that you have what you want!

JG: He wouldn't even bother to fight with you, to get into power struggles with you, unless he thought of you as a worthy adversary.

Harry: (Laughter.) Yeah!

RLP: Now, Harry, we have to draw this to a close. Let me give you my impressions. We'll type them up and send them to you, and you'll make use of them however you want to make use of them. Before I state them, though, I want to ask you how it feels for you to have had this conversation with us? We only met today, and here we've gotten to talk about all kinds of personal things.

Harry: It's good. It feels good. It's interesting. I'm hoping I've gotten something that I'm going to turn into some positive stimulus for myself, for how I see things. There are things I want to sit down and think about now.

RLP: Good. Let's see if I can come up with some impressions that will be useful to you.

[Note: At this point RLP made the following statement, in Harry's presence.   JG   wrote   the   statement   down,   pausing   from time to time to enter into further conversation with RLP about a choice of words, or a detail which had been forgotten, etc.]

The Case of Harry (continued)

Summary of Impressions Derived from the Initial Interview

Harry presents himself at what to him is the "magic" age of thirty: The time in a man's life when he should be able to demonstrate his success, consummate a good marriage, and be the head of a family. He should no longer be acting like a young fellow in his twenties, unsure of himself, unsure of his goals, and unmarried.

He has, therefore, set out to do what he should do: Get a wife, start a family, get his own career independent of the family business, and demonstrate to his father that he is entitled to respect as a man amongst men.

Unfortunately, it has not yet clearly occurred to him that he requires the cooperation of an enthusiastic partner to make a marriage and a family. Furthermore, until now he has not seen clearly that the two goals of establishing himself in a career and impressing his father are mutually exclusive.

That is to say, if he wants above all else to impress his father favorably, he must be prepared to act the way Sidney does in the business. If he wants to establish an independent career for himself, he has to do what Father did in his life, namely, set his own goals and pursue his own course, disregarding anyone else's opinion as to the wisdom or desirability of his choices.

[Note: A method for composing the *Summary of Impressions Derived from the Initial Interview* is presented in Chapter 12.]

Conclusion

Our intention in presenting the complete discussion between Harry and the two therapists is to demonstrate that a person who appears to have several different concerns can gain a unifying clarity in the space of a single interview. Use of the *IPCW* inquiry format, combined with client and therapists working together in the review process, can lead to an understanding of what the client is doing, what he or she is doing it for, what it costs to keep it going, and what else is open to his or her choice.

The remaining chapters in Part III will guide the reader through the review and interpretations of the family constellation data, revealing how key features of the client's childhood experience provide a context and background for understanding the person's interpretations of the current situation.

Chapter 6. GENDER GUIDING LINES, ROLE MODELS, AND THE BIG NUMBERS

The image of fictional guiding lines was used as an early theoretical construct in the development of Individual Psychology, a construct later included among ideas represented by the term lifestyle. As meridians on a chart facilitate navigation across the surface of the earth, so guiding lines provide orientation for individual movement in the social world.

It is important to state that theories regarding the origins of sexual identity, sexual orientation, and sexual practices are not at issue here; our focus is the unique style of living presented by the individual variant. Individual Psychology has always stressed the fact that it regards psychic life as *movement*, and considers form, expressions, function as a kind of frozen movement (Adler, 1970, p. 118). We think of the style of living, therefore, by analogy to handwriting, generally considered unique.

Although we expect methods of inquiry outlined in this chapter to assist clinicians in gaining an understanding of a client's movement, including behaviors associated with sex and gender, we explicitly reject such terms as "homosexual lifestyle" or "heterosexual lifestyle" as these suggest a necessary connection between sexual orientation and other values and attitudes practiced by any member of an identified sub-group or category of affiliation. Such notions contradict the Adlerian understanding of each person as a unique variant of human possibility and expression.

It is also important to set forth our understanding of the terms, sex and gender. *Sex* refers specifically to biological *structures*. Male and female are *sexes*, denoted as such by their unambiguous biological structures. There are also the *intersexes*, whose sexual structures are mixed. The developmental geneticist Anne Fausto-Sterling (1993) states the "two-party sexual system" is "in defiance of nature" as "there are many gradations running from female to male" and that "one can argue that along that spectrum there are at least five sexes" (p. 21).

Gender and *gender identity*, on the other hand, have to do with *meaning*. These terms do not relate to structures, but to *constructions*, both personal and social. Gender refers to the ways individuals evaluate and give meanings to their sexual structures and to the ways they direct their sexual feelings, both in the personal and social realms, according to how they have shaped or are shaping these matters for themselves. In Adlerian Psychology, nothing is *determined*; rather, as Adler (1980) stated it, "Meanings are not determined by situations, but we determine ourselves by the meanings we give to situations" (p. 14).

The gender guiding lines are those evaluations and expectations a child employs in forming a schema of apperception. This schema, formed early in life and unique to the individual, provides the axiomatic foundation from which the reasoning of the private sense proceeds.

Since subsequent impressions are filtered through this schema, the biases of the preverbal and early verbal period carry through, with more or less modification, into adulthood. As Adler put it, these impressions or guiding lines are "attached to the individual as a pattern, permitting him to express his self-consistent personality in any situation without much reflection" (p. 219).

Among the most important and ineluctable of the early impressions is the picture the child forms of what it means to be a man or to be a woman; what it means to be one of these among others of the same and of the other sex; and, what being a man or being a woman means with respect to the struggle to overcome inferiorities and to meet the tasks life imposes.

Success in life depends, in part, upon our ability to solve the problems connected with sexuality. Apart from the anomalies associated with intersex (where matters may be more indeterminate, and sexual recognition may be delayed or imposed), each of us is recognized at birth as a member of one sex or the other. In either case, we encounter the problems of sexuality in a particular cultural context of meanings. Reproductive love and the securing of children in the supportive bonds of family life are tasks set for two persons (Adler, 1982, p. 124). To succeed in these tasks, therefore, even to be able to address them at all, requires that we develop a readiness for cooperation with a sexual partner.

We cannot depend on fixed action patterns, as the other animals appear to do, for the solution of the problems associated with these tasks. Instead, we have to rely on the symbols and images we ourselves create, using our own ideas about what it means to be male or female, to govern our sex-related attitudes, convictions, expectations, and operations.

At each stage of our development throughout childhood and adolescence we operate as if we were asking and rehearsing trial answers to questions such as, "What does my gender require of me, what does it promise me, and what am I going to do about it in order to secure a place (or avoid a place) with a sexual partner?"

When gender definition is conveyed to us through images of masculine and feminine that are both desirable to us and useful in our social world, we feel at ease in our sexuality, and in our relations with other members of our sex as well as of the other sex.

If these images are mistaken, particularly with respect to the demand for cooperation, and if they exaggerate either the requirements or the promises of our gender's social role, we may experience trouble in our sexual dealings. Then our capacities for sexual functioning, for connecting with a partner in love and intimacy, and for success as one of two cooperating parents, will be undermined, whatever our sexual orientation may be.

In our present historical circumstances cultural images of the differences between the sexes (and of the proper biological roles that each has to perform to ensure the future of the species) remain distorted by an autocratic tradition of inequality. Images of masculine and feminine are therefore often unattractive (in that they point backward to standards that are no longer acceptable) or unclear (in that they point forward to standards that are not yet culturally defined). Many individuals are burdened with hesitation in their confusion about what it means to be persons of their sex, and about what may be required of them to engage intimately (and happily) with another.

Adler understood the importance of these questions in their effects on basic feelings of dignity. His awareness of the damaging effect of a social and symbolic system defining the sexes in terms of inequality, led him to formulate the concept of "the masculine protest." Since American English has largely lost the original meaning of "protest," this has become a less and less useful phrase, easily mistaken for an objection or a complaint.

These meanings can be included in "protest," which is, however, properly understood as an assertion or a claim made by either a woman or a man for the (perceived) male position of dignity and status with its (perceived) prerogatives and power. Adler posited that the girl or woman senses her socially inferior position to the boy or man and strives to feel equal. Her goal: Treat me like a man! In the event, she feels equal to the man she perceives as *superior* to her *only* if she senses herself to be "equally" superior to the man. Adler also saw the boy and man who perceive themselves to be inferior to other men (or, worse, to girls and women) protesting: Treat me like a Real Man!

To understand a lifestyle, and especially to search out and identify the errors standing in the way of successful adaptation, it is, therefore, of central importance to understand the person's governing images or guiding lines relating to gender. We are also alert to gender exemplars the individual has chosen either to emulate (positive role models) or to attend to as cautionary examples (negative role models).

As a general rule, the more disturbed and misfitted the individual is by reason of a faulty style, the more we can expect that person to interpret sexuality (and all gender-related imagery) in oppositional terms. Adler therefore recommended that the very phrase, the opposite sex, be abandoned, as it encourages a continuation of such ideas, and called, instead, for use of the phrase, the other sex. More and more often we can see a direct correlation between the extent of dysfunctional movement and the intensity of ideas of inequality between the sexes. As Adler put it:

In the [private] sense of the patient, but not in the general [common]sense, memories, impulses, and actions are always arranged according to a classification of *inferior=below=feminine* versus *powerful=above=masculine*. (p. 249)

Dreikurs used Adler's general concept of guiding lines to identify a person's governing images of masculine and feminine, and what it means to the person to be a man or a woman. In our continuing exploration of the importance of these images, we have discovered that it is greatly to our advantage to maintain a distinction between role models and gender guiding lines as theoretical constructs. (See Figures 5 and 6, below, pp. 188-189.)

ROLE MODELS

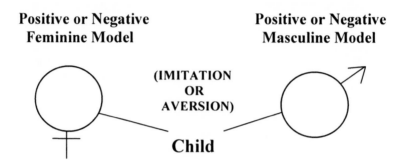

Positive or Negative
Feminine Model

Positive or Negative
Masculine Model

(IMITATION
OR
AVERSION)

Child

1. Solid lines represent the ***conscious*** assessment the child makes of individuals regarded as impressive, either positively or negatively.
2. The child chooses these individuals as role models, thinking, "When I grow up I want/don't want to be like that."
3. The child experiences itself as desiring to move toward/away from its role models without the sense of destiny felt regarding the gender guiding lines.
4. When the parent of the *same sex* is a *positive* role model, chosen in awareness and consciously imitated, one is likely to feel consonance and success in gender-related behavior.
5. When the parent of the *same sex* is chosen as a *negative* role model, one is more likely to feel dissonance and unease in gender-related behavior.
6. Similarly, when the parent of the *other sex* is chosen as a *positive* role model, dissonance and unease is more likely to be felt in gender-related behavior.

Figure 5.

GENDER GUIDING LINES

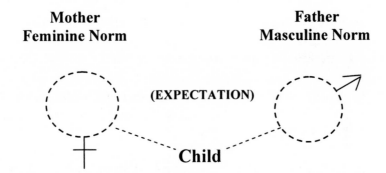

1. Dotted lines represent the **unconscious** quality of the child's sense of what it means to be a man or a woman. The child, without conscious awareness, is forming opinions about gender attributes. These convictions, assumptions, and expectations are the basis of patterns of thought and behavior that usually remain unexamined in subsequent adult life.
2. The child experiences itself as destined to be a person like the same-sex parent, *as if* thinking, "When I grow up to be a man/woman, I'm going to be like that — unless I *do* something to change it."
3. Mother and Father (or parental substitutes in the absence of parents) are accepted *as if* setting the norm to which other men and women either conform or from which they deviate, as in a Bell curve.

Figure 6.

Gender Guiding Lines

We understand the gender guiding lines to be:

1. *largely outside* the person's conscious awareness;

2. patterns for the person's *expectations* regarding gender;

3. experienced as if presenting a *destiny*; and

4. shaped in conformity with *images* of mother and father (or parental substitutes), serving as the *norms* for what a real man and a real woman are.

According to this understanding, children form their pictures of what it means to be a man or a woman by experiencing, and assigning meanings to, the differences between mother and father. When we were children, our images of mother and father established the norms for what it is to be a man or to be a woman. Without having to think about it, we compared other men to father, other women to mother, and found that the others deviated in one or more particulars from these norms, either by going beyond expectations dictated by the guiding lines, or by falling short of them. It is as if we were noting, "All men are like my father or unlike my father; all women are like my mother or unlike my mother."

This process takes place against a background of cultural norms and values. It therefore requires little, if any, conscious awareness. The assumptions formed by a child in the early years of life regarding his or her own sex and the other sex continue throughout life, largely unexamined, to the extent that they remain effective for the individual (however troublesome or offensive they may appear from others' points of view).

These assumptions become the child's expectations concerning gender. Contrived in the universal striving to overcome feelings of inferiority, the guiding lines point the way to a vision of the future, in which weak, insecure, and dependent children can picture themselves enjoying the strong, secure, and independent status of adults. Bringing an image of future competence into the frame of reference of the present phenomenological field, the guiding line functions as a compensatory mechanism in the present, as a source of encouragement, and as a template for training and self-training.

Children shape images of themselves as adults along the guiding lines, which they come to regard as if tracing a destiny. As the capacities for critical discrimination grow, and the guiding lines are seen to encompass unattractive, unwelcome, or repugnant features, the sense of destiny may reveal itself in an uneasiness, as if the child were thinking, "When I grow up I'm going to be like that (referring to qualities of the same-sex parent), unless I do something about it."

Even in situations where one (or both) of the parents is absent, the child forms an image of the missing parent(s). A parent may be absent owing to death, divorce, abandonment, confinement to an institution, or other circumstances. One must consider situations in which a child is raised entirely by a single parent or by an adoptive or foster single parent, or by a single grandparent, aunt, or uncle, or other single caretaker. In any of these cases, a child will seize upon scraps of available information or even speculation (such as may be derived from a photograph, overheard bits of conversation among relatives, or personal possessions of the actual but missing parent) and organize the impressions gained from these against the "whole cloth" of creative imagination. The resulting picture serves as the gender guiding line set by that parent.

Case example. Author Susan Sontag is reported (Nunez, 2011) to have lost her father to tuberculosis when she was five years old. "Because she barely remembered her father and was able to learn little about him, she had had to invent him. . . . She imagined her father, though he had not been highly educated, endowed with a good mind and other qualities she could admire. . . . She liked to think that, had he lived, he would have been a good father to her, the one family member she would have been able to relate to, proud of her achievements, able to share her enthusiasms" (pp. 27-28).

Case example. Margie's parents divorced when she was two years old. She retained no memories of her natural father. Her mother remarried shortly after the divorce, and Margie grew up with an affectionate stepfather whom she called Father, the only father she had any sense of ever having known. Stepfather was quiet, unobtrusive, a steady person who was a hard worker and a good provider. Margie did not see her natural father, nor was he in touch with her in any way until she was in her thirties.

At twenty-two, Margie married a man whom she described as "flashy, impulsive, passionate, and exciting." In short, he was in every way as different from her stepfather as could be imagined. In an exploration of gender guiding lines she said, "I don't know how I could have ever married a man like Tommy." We asked if she knew anything about her natural father. She said that all she really knew was what she had overheard once when she was eight or nine years old in a conversation between Mother and Grandmother in which they were talking about his gambling habit and his fast-living style. She had been intrigued by this. Then, there were a few occasions when she knew that he was visiting in town, because there had been conversations between Mother and other people as to when "Ralph" was expected to arrive. She remembered comments such as, "I sure hope I get to see Ralph when he's here! Boy, is he a character!" Mother and Grandmother were abuzz at these times, and people Margie did not even know would speak to her mother on the street, saying, "I hear Ralph's going to be in town — he'll put on a show!"

Margie never saw her father during these visits. Secretly, however, she wove an image of what this dashing, entertaining man was like, and mooned over it, thinking how grand it would be if he would come and sweep her away from the life she was leading, which, in comparison with her fantasy, was humdrum and colorless. As she told the story, she began to smile with recognition, and, displaying a pronounced blush, said, "Now I understand why I chose Tommy!" She realized Tommy had met her expectations for what a "real man" was.

Role Models

We distinguish role models from the gender guiding lines, and understand them to be:

1. largely *in conscious awareness* throughout childhood and adolescence;

2. patterns for the person's *imitation* or *aversion*, independent of gender;

3. experienced as if providing an escape from destiny by way of *choice*; and

4. shaped in conformity with images of men and women who represent possibilities for *deviation from the norms.*

Children, ineluctably and increasingly aware of the importance of gender, attend to and study both its positive and negative possibilities. In the early years of children's development they encountered many men and women other than parents. Grandparents and other relatives, teachers, preachers, bus drivers, letter carriers, neighbors, and the parents of other children all played parts in their world. Some of these people took on a special importance as illustrating something about human possibilities, and therefore about gender possibilities. We hoped or wished to be like some of them; we wanted to avoid being like some others.

Expectations characterize the guiding lines; the parallel terms for the role models are imitation and aversion. We want our lives to be like those we admire, and we study them in imitation. These are our *positive role models.* We deliberately strive to copy those aspects of their personalities and histories that appeal to us. In the same way, we set out deliberately to avoid being like people we regard as reprehensible, offensive, or unattractive. These are our *negative role models*, and we study them in an effort to reject the possibility of behaving in any way that reminds us of them.

We ask clients to tell us about any adults other than the parents who were important to them when they were growing up. Usually two or three come to mind.

Case example. Edith described her mother as a perfectionistic, fault-finding woman whom Edith never pleased no matter how hard she tried. Mother seemed to Edith to complain about everything, and not to enjoy life, except with respect to new clothes, and especially when she was able to dress for a social occasion: Since she had a striking face and figure, she could do this very well indeed, and was much admired by Father (and others) for her style and flair. Mother's sister, on the other hand, was a happy and accepting person. "She'd just let me be me. She liked me and was always glad to see me. We spent a lot of time together, and she'd always be cooking up a storm — good things. Mother wasn't interested in that at all. Aunt Carolyn was fun! I wanted to be like her, not like my mother."

When a child finds the behavior of the same-sex parent to be unattractive, as Edith did, she will regard that parent as a negative role model, and set herself the task of avoiding any imitation of it. She will look for positive role models as encouraging examples of what else is open to her as a girl who will grow up to be a woman. Edith's Aunt Carolyn was one of these for her.

It is usual, however, that such a person will feel some sense of uneasiness resulting from the struggle she is having to free herself from the gender expectations dictated by the guiding line set by the parent of the same sex. In Edith's case she safeguarded in an exaggerated way against criticizing anyone or complaining against anything for fear of becoming a fault-finding woman like her mother. She had come to find it hard to let others know where she stood, for fear of appearing critical. She could also feel tormented by her sense of being critical of people who are critical, and, in a kind of psychic hall of mirrors, found herself trying to be perfectly free of perfectionism!

Edith was also a woman of striking good looks, like her mother, and this likeness felt to her to be another danger, which she fought against by a studied avoidance of anything fashionable. Even this did not keep her safe, however; the studied character of her avoidance resulted in her having a stylishness of her own that people found attractive. Finally, in deliberate imitation of Aunt Carolyn she had studied and mastered the culinary arts, and continued to enjoy "cooking up a storm" — except on those occasions when something went wrong with a recipe: "You see," she'd say to herself or to anyone else who would listen, "I'm just like my mother when it comes to cooking."

The sense of dissonance and uneasiness increases for the child who not only decides not to be like the same-sex parent, but who also sets out to imitate the parent of the other sex. In such a case there is a conflict between one's gender and the chosen role model, and the person's experience of gender identity may be clouded by a suspicion of inauthenticity. It is as if the person were asking, "If I am like my mother, what kind of a man am I? Am I a real man?" Or, "If I am like my father, what kind of a woman am I? Am I a real woman?"

Case example. Frank reported that his father was a bully and a showoff. He beat the children, and was always getting into fights and being suspended from work. He abused Mother as well. This ended when, at sixteen years of age, Frank intervened in a fight between Mother and Father. He knocked his father down and said, "Don't ever hit her again, or I'll throw you out." Father knew he was beaten, and timidly acquiesced. Frank was pleased to have ended the violence between Mother and Father, but was uneasy over the humiliation he had inflicted on his father through his own "brutality."

Frank had always seen his mother as a kind, fair-minded person who worked hard and did what she could to maintain peace in the family. He admired her. He grew up thinking, "I don't want to be like my father. I want to be like Mother, a person of character."

As an adult, Frank is soft-spoken and low-keyed. He is a hard worker and a family man. He is also uneasy about what kind of man he could possibly be, and troubled by fears that he may be "effeminate." He won't fight to get his own way, least of all by arguing with his wife. He says, "At work I let people walk all over me, and at home the kids just run wild when I'm alone with them." He is preoccupied with a fear that, "Really, down deep, I may be gay." He asks, "What's wrong with me?"

Frank's adult functioning is constricted by the safeguard he has erected against the "destiny" of emerging as a brute, which, he believed, could occur at any time he were so much as to express himself in a forceful tone. He would then be "revealing himself" in the image defined by his masculine guiding line, and would be seen as the (dreaded) "real man" of his expectation.

An individual can experience another difficulty of this kind when a positive role model is recommended to the child by a parent, who offers it as a standard of personal value. As an example, the parent of the other sex says (or, more likely, implies) to the child, "Don't be like *your* mother; be like *my* mother" (father to daughter), or "Don't be like *your* father; be like *my* father" (mother to son).

Case example. Judy saw her mother as a good-hearted and well-intentioned person. But, Judy said, "Mother was confused. She just couldn't get organized. She seemed to be in a private world. It was a daily problem. For example, she never could get dinner together and ready on time. Dad would come home on a tight schedule (and he didn't demand very much from her, as it appeared to me), and dinner wouldn't be ready. This didn't happen once; it happened night, after night, after night. She was late to every appointment she ever had. God, how they'd argue and fight over this. And it was hopeless! He couldn't count on her for *anything*. It was so different at my grandmother's (his mother's) house. Everything was so orderly there. Dad was always comparing my mother to his mother, and my mother always came out on the short end of the stick. I don't know why he stayed with her. Well . . . she was the prettiest woman in our town, and she was so vain! I guess that had a lot to do with it. Anyway, I never wanted to be like my mother. So I'm a capable, orderly person — but I don't know how 'feminine' I am, or whether I'd get the attention of a real man like my father. Mother was such a bubblebrain, but she had him! I'm so down to earth and practical that it scares me."

There would be no reason for Judy's worried appraisal of her everyday sensible and practical approach to things unless she had a basic idea that a "real" woman (that is to say, a woman capable of attracting and holding a "real" man) pays the price for her advantages in a loss of dignity: incompetent, self-absorbed, and flighty, like Mother.

All compensation tends toward overcompensation, and Judy was so safeguarded against being a "bubblebrain" that she was rigid, lacking spontaneity and lightness in her approach to life. She was following the role model presented by Grandmother to compensate against the danger of feminine inferiority represented in the guiding line she saw set by her own mother. She had lost sight of her own femininity (and her grandmother's), experiencing a *dissonance* between her gender guiding line (which had its appeal, even to her) and her role model, to which she felt (so to speak) "honor bound," as a way of maintaining her self-esteem. Until she explored these issues in therapy she was uneasy in her self-appraisal of her attractiveness and sexuality, as if she believed, "I'm not truly feminine."

The sense of expectation that flows from the gender guiding lines dictates attitudes toward one's own, and toward the other, sex. Those who conform to our hidden gender guiding lines (even though we may not like them), remain recognizable to us as "the genuine article," either as men or as women.

We sometimes express our gender guiding lines by stumbling over them, even though, typically, we remain unaware of what we express in our unguarded outbursts. For example, when a man (or a woman) says, "Women are like that!" or when a woman (and sometimes a man) says, "Isn't that just like a man!" some basic assumptions about gender are being expressed. These expectations for one's own and the other sex predicate the norms of personal movement in relation to both sexes.

Case example. Jim grew up in a household in which Mother was bipolar. His sister was a sickly child, and one of his aunts, who lived nearby, was an alcoholic. He had concluded: "You can't count on a woman; either they're screwy or they're sick. In either case, the men have to take care of them." His feminine guiding line, established by his mother's wild flights alternating with silent withdrawals, was reinforced by the periodic illness of his sister and his aunt's episodic bouts of intoxication. The principle he derived from this was: Whatever else is true about them, women are not to be trusted.

Jim's picture of Father, by contrast, was all positive. He saw Father as steady in temperament, tolerant, and unfailingly reliable as a caretaker and provider. Jim loved Father, and was able to count on him for a secure sense of having a place in the world. Father never defaulted; he endured.

Father provided Jim with a noble masculine guiding line and, since Jim wanted nothing else so much as to be like Father, Jim had self-consciously chosen Father as a positive role model as well. Here there was a clear case of *consonance* between the destiny of the guiding line and the open desire to pursue the role model. Ordinarily, in such a case, we can expect the result to be a comfortable adaptation to adult gender functioning and sexuality.

In Jim's case the life of a man seemed to be a tremendous challenge and a heroic burden. He had grown up admiring his loyal and reliable father, and at the same time wondering, "Do I have what it takes to be like that?" He set out to prove that he did, in order to succeed in the task of being a "real" man.

Not surprisingly, the only women who managed to catch and hold his attention were those who appeared to require a caretaker. When he came to see us, it was at the end of four years in a relationship with a woman who couldn't hold a job because of a variety of somatic symptoms, who had constant money troubles, and who couldn't maintain friendships because of her hair-trigger temper. Her first husband had abandoned her, and she was raising a son who (understandably) adored Jim. Jim found it impossible to break off with her (and her son) in spite of her endless complaints, demands, and fits of temper. After all, his motto was, "Men don't give up!" If he were to give up on her, he wondered, "What kind of man would I be?" His heroic picture of what is required of a man had kept him in a destructive relationship with a woman who was prepared to exploit him.

As with all preparations for life, the gender guiding lines can be traced to efforts at orientation begun in early childhood. As with all human phenomena, this does not require us to conclude that they were irremediably fixed in childhood. Their fluidity remains especially obvious through adolescence, when they are still subject to dramatic modification on the basis of new experiences and information.

Case example. Mark told us that until he was eleven years old he had seen his father as flawless. Father doted on his wife and children. A successful businessman, he was well-respected in the community, gregarious, and an athlete who enjoyed a Saturday afternoon soccer game. When Mark was eleven he learned accidentally that Father was involved in a long-standing affair with his secretary. Mark's world seemed to collapse. From that point on he thought, "No matter how men look and act, the hidden truth is that they can betray everything important in life." Mark saw his mother as a good and kind woman, an encouraging person, and fun to be around. He saw her as the victim of his father's misbehavior. He hated his father for what he had done to "this good woman."

How did these ideas express themselves in the movement of Mark's adult life?   His first sexual affair was with a married woman, and his next liaison went on over the course of several years, during which his partner deceived him with other men numerous times. His private sense was: Since I am a man, I cannot be counted on to be faithful to a woman. I must therefore not allow myself to become involved with a good woman whom I would ultimately betray. Therefore, I must find a woman who is worse than I am, who will keep me safe against the shame of being a traitor by being the one who will betray me.

Further Examples

The Case of Catherine. Catherine saw Father as cheerful, confident, and competent. He was also playful and uncritical. He was at home in the world, presenting himself as a friendly person. She saw Mother as fretful, concerned with propriety, ill-at-ease amongst others, and dependent upon Father to make everything in life both work and work out. Catherine's gender guiding lines, therefore, presented men as the enablers of life; and women as dependent upon the men for whatever the men chose to do or not to do for them. Women were basically misfitted to do anything but to look nice, and to be fretful if things did not go as they would have liked, according to Catherine's view. The only power she clearly saw as belonging to women was that of moral judgment. Catherine, believing that everything is up to the men, has been waiting for a man to make life for her. Until then, she concerns herself with her appearance, with biding her time, and with being critical of others.

Until these issues were explored, Catherine had not seen herself as an actor who, by what she does and does not do, has an effect upon those around her.

The Case of Roger. Roger saw Father as encouraging, outgoing, and practical; Mother was righteous, quarrelsome, and suffering. No matter how hard Roger and Father tried to please Mother, they were never able to succeed. It seemed that all their efforts only served to increase Mother's sense that life had been unfair to her, and to invite her further complaints against one and all for how much she had been, and still was being, mistreated.

Roger expects a woman to suffer, and that it will be his manly duty to make it up to her for what she suffers. The only way to avoid this position is not to get involved with a woman; or, failing that, to keep one's distance and to protect one's privacy. Roger has a job that takes him on the road all week, away from the suffering woman to whom he has been married for the past fifteen unhappy years.

The Case of Joe. Joe saw Father as a hard worker, a man who could overpower others by a tyrannical style of demands and expectations no one could meet and no one could satisfy. He saw Mother as caring and cautious, and as encouraging the children to better themselves as a way of escaping Father's domination. Joe has a good alliance with his wife, but has trouble with members of his own sex. He expects it to be "a jungle out there" in the world of men, especially those men who are in positions senior to his in the business world. He believes that they will demand too much, that their expectations will be too high, and that, since there will be no way for him to please them, his only hope is to escape them. Joe had changed jobs many times before he decided to explore this issue in therapy.

The Case of Julie. Julie saw Father as wonderful. He worked hard, was reliable, and was uncomplaining. Although he did not encourage her, he never hampered Julie in what she wanted to do. Father was practical, and led a life of routine.

Julie saw Mother as restless, complaining, and determined to do whatever it was she wanted to do, even if it was beyond the family's means, and even if Father objected. Mother was frivolous and selfish, according to Julie. As an adult, Julie believes herself to be tempted to be self-centered, and is determined not to yield to this temptation.

She does not experience herself as having anything to offer to a man. In her view men are wonderful, and women are a drag on them. She cannot go out with "good" men, therefore, since she believes that she would only mess things up for them and hurt them. Instead, she dates a series of unsuitable partners who get bored with her and leave her.

She works very well with men in the business setting, but expressed her bias against her own sex when she said that she never gets along with the women there, and that, "They just do whatever they want to do without any thought of anyone else. It's infuriating!"

The Case of Rafael.   Rafael grew up in a Puerto Rican neighborhood that was close-knit and full of his relatives. He was the only child of his parents, whose ambition to have a large family had been thwarted: When Rafael was two years old Father had suffered an injury that left him invalided for life. Mother became the breadwinner and the caretaker at that time, and Rafael grew up feeling sorry for her. She was a good woman, and she deserved better.

Members of the extended family made a pet of Rafael, who spent as much time with them as possible. At home, Father became increasingly tyrannical and embittered, managing to rule both Mother and Rafael from his bed by means of temper outbursts, insults, and accusations. Mother acquiesced as if to her fate; Rafael obeyed while maintaining his defiance. He came to hate his father.

The other men in his family did not offer attractive models: One of his grandfathers was known as a "drifter" who moved from town to town, leaving Grandmother to care for the household as best she could. He would return periodically, to be nursed back to health, after which he would leave again. The other grandfather, while a steady and kindly man, was a "dreamer," interested in woodworking, in nature, music, and (according to Mother) "things that don't matter," and both Mother and Grandmother dismissed him as foolish and inconsequential. The two uncles Rafael knew best couldn't hold jobs. It seemed to Rafael that there was no good way to be a man. Besides this, he liked his position as a little boy, fussed over and enjoyed. He developed and practiced a gay identity, *as if* to proclaim: I would rather identify with the female side. Men are defaulters, either troublesome or silly. Women, on the other hand, hold the world together with responsible caretaking and practical skills. I don't *want* to grow up at all, but if I must, I will take the female part, which at least has some dignity attached to it.

He came to see us complaining that his lover, a self-anointed person of importance in the theater, was exploiting him under the guise of helping him in his career. The lover's behavior became more and more bizarre in the following weeks, coincidental with Rafael's work with us in a lifestyle assessment. Rafael felt responsible to take care of his lover as things deteriorated, but was unable to manage the situation as it eventuated in the lover's having a psychotic episode requiring admission to a hospital, leaving Rafael stuck with the lease and the obligation to pay the rent on their apartment.

The Big Numbers

Gender Guiding Lines and Age Expectations

In our clinical practice we have further noticed that the ages at which particular events took place in the life of a client's same-sex parent can serve as psychological "markers" or points of reference, forming a private timetable against which to measure personal progress, for better or worse.

We call these the "Big Numbers," a phrase we have chosen to represent the power of these convictions about how life can be expected to unfold. We think of these as analogous to the age-markers sometimes regarded as of special importance in the commonsense of the culture, such as "life begins at forty." Here, however, we are interested in numbers regarding age which are peculiar to the private sense of an individual.

Since *expectations are the primary motivators*, obscure and unrecognized images will influence the course of a person's movement in sometimes puzzling ways — until they are brought into awareness and reexamined.

Case example. Carl, thirty-seven, complained of loss of enthusiasm. He said that for the past year he had had no sense of direction, and that he was afraid he was "going crazy." We began an inquiry into his family background. As soon as we had learned about his father's place in the constellation, we put down our pencils and said, "You're not going crazy, and it doesn't appear that you need psychotherapy. You need to work out what we call an 'Alternate Plan B,' for use in case of success." He looked at us quizzically, so we went on: "You have already explained the matter to us. Your father died suddenly when you were twelve years old, leaving no money. As the firstborn, it was up to you to help Mother survive. You worked hard, put yourself through school, and helped your younger brother and sister through as well. You told us earlier that you have worked hard in your profession, and used your real estate investments to secure the future for your wife and children. The only problem is that you didn't die last year as you were supposed to, leaving the world to remember you reverently and fondly." He grinned in recognition. We said: "Your father was twenty-four when you were born. So he was thirty-six when he died, leaving you to carry the whole weight of the family. Our guess is that you swore that, whatever else happened in your life, you would never do what he did. And you won't. But there's one thing missing: You failed to die when you were supposed to." He was blushing now, as well as grinning, and tears filled his eyes. "I never would have said it — but it feels as if you've read my mind! So. All I need is a 'Plan B'!" Later he called to tell us that he felt no need for further discussions, adding, "My wife and I are busy making plans for what to do in case we go on living! We've never been happier."

Case example. Joan, a personable, bright, and well-educated woman, complained of trouble at work where she was "acting bizarre," "blowing up" at her boss, offending clients and co-workers. She had received notice from her company. She had also been fired two years before, by her previous employer. She was thirty-two years old, and her next birthday was five weeks away. She had been married for four years, and said that it was a good marriage except that her husband was stepping up the pressure for her to get pregnant and start a family, and that, "I don't want to get pregnant until my career is on track." A brief inquiry into family background revealed that Joan was the firstborn child of her mother. Mother was thirty-two when Joan was born, thirty-three when she had her second child, and had given birth to her third and last child only eleven months later. Mother never returned to her career, though she talked about it endlessly, threatening the children and complaining, "I'm going to leave you all! You've ruined my career! You're in the way!" Joan could appeal to the current commonsense in stating that she was eager to have children as soon as her career was sufficiently developed, so that she could return to it later. In Joan's private sense, however, motherhood necessitated a miserable life of service and self-sacrifice. Her disturbing behavior at work had begun as she was approaching her thirty-first birthday (private sense = first age at which to conceive), and was now culminating as she began to approach her thirty-fourth year (private sense = last age at which to conceive). It stood in sharp contrast to her earlier record of competence and success. It had been serving a motive, until now "not-understood" (Adler, p. 192): It provided an "honorable way out" of the parenthood she claimed to desire, and an escape from an expected defeat.

In these two examples the Big Numbers related to negative expectations and pessimistic evasion; however, the Big Numbers may also relate to positive expectations, expressed in overambitious hesitation. Another man came to see us at age thirty-six. By the time his father was thirty-six, Father had been made partner in a prestigious law firm, had married, built a house, and welcomed the birth of his fourth child. Our client's achievements paled, in his mind, in comparison to his father's record. Unable to meet his (hidden) timetable, he said, "I am such a failure. I feel like my life is over."

Individual evaluations of the "destiny" may undermine or encourage, and even a negative image may stimulate beneficial compensations. Paul A. Samuelson, the first American recipient of the Nobel Prize in Economics, said:

> Consciously or unconsciously, I was a young man in a hurry because I felt that the limited lifespan of my male ancestors tolled the knell for me. My father died when I was twenty-three. I was supposed to resemble him, and the effect on me was traumatic. What I was to do I would have to do early, I thought. Actually, modern science granted me a respite. . . . Whatever the reason, I have been granted bounding good health. (Breit & Hirsch, 2004, p. 49)

Individuals often have private timetables based not only on the lives of their same-sex parents, but also on the lives of their parents of the other sex. As with Big Numbers and the ways in which individuals expect their own lives to unfold, they also form expectations for the ways in which the *lives of their partners* will unfold. If a client's father died at an early age of heart disease, she may have an expectation that her husband is at risk of dying in the same way unless she does something about it. She may then behave in overprotective ways that her spouse resents as controlling or condescending, and may thereby introduce trouble into the relationship until the underlying conviction is understood and reconsidered.

It also appears the Big Numbers can be significant in the relationships of parents with their children. For example, they may expect their same-sex children to falter or make mistakes at certain ages, analogous to difficulties or errors they remember having made in the course of their own development. (This is even more likely to occur if parent and child share not only gender, but also birthorder position.)

Three Anecdotes

One person reported: "My eyes were opened when I learned about the Big Numbers. In my case, I had never expected to live beyond age twenty-eight, the age at which my mother died. I now realize for the first time that at that age I completely changed my everyday life when I went back to school full-time for the following five years. I had always assumed that I was a 'late bloomer,' not knowing that I had been waiting to see what would happen at twenty-eight. Looking back now, twenty years later, I can see that the Big Numbers really mean something."

Another client, who was approaching age thirty, said that she never expected to live to that age and now felt time was running out on her opportunity to start a family. She connected these ideas and feelings to the fact that her mother was thirty when she (the younger of two) was born, and that her mother had then immediately entered menopause.

A student wrote, "My mom had a difficult time turning forty-eight because that was how old her own mother was when she died. No matter what we did to convince my mom that she wouldn't die at forty-eight, it did not work. On her forty-eighth birthday, when she woke up and realized that she was still alive, Mom decided to buy an organ and take music lessons, something she had always wanted to do but never did until then."

Research Reports

Research based on a study of 2,745,149 deaths from natural causes demonstrates sex differences in what the authors call "anniversary effects," which they define as behavior changes that take place around significant dates in the life of an individual. The study, published in *Psychosomatic Medicine* (12/92), reports that women are more likely to die *after* birthdays, while men are more likely to die *before* birthdays.

David B. Phillips, Ph.D., a sociologist at the University of California – San Diego, who conducted the study, uses the terms "lifeline" and "deadline" to characterize his explanatory hypotheses, which is that for women birthdays are a lifeline because they are times to enjoy family and other social relationships and to receive special attention, while for men birthdays are a deadline because they represent a time for taking stock on one's life and accomplishments. He makes these guesses based upon his sense of the cultural socialization of women for social success, and of men for achievement.

Phillips disclaims any understanding of the biology that sustains or shortens life in these situations, but acknowledges, with other researchers, the impact of symbolic events on general health.

Even though this research addresses statistical phenomena in large populations, and we are interested in what Adler called the individual variant and idiosyncratic meanings, it is, nonetheless, reliable evidence in support of Big Numbers theory.

One further note. *Newsweek*, in a combined "Special Election Issue" (11/92 & 12/92), provided this report on William J. Clinton, then President-elect:

> The death of his father in a car crash months before he was born left its mark on him. In philosophical moments [during the campaign] he would say, "I always felt I wouldn't have enough time."

Conclusion

The basis for present gender-related attitudes and operations of the client can become clear in a review of the client's gender guiding lines and role models. Sometimes it is possible to relate lifestyle data recorded in the family material to responses to the questions in the initial interview, "What makes a man 'masculine' to you? What makes a woman 'feminine' to you?" Correlations between mother's and father's and role models' attractive and unattractive features, as sketched out by the client in the first session, provide dramatic confirmation of the self-consistent character of an individual's attitudes and values.

Following upon this review of gender-related values, we will direct our attention toward an examination of the atmosphere created by the parents (and others in the household), the values they communicated, and the conclusions reached by the child concerning interpersonal relations and a personal code of behavior. These topics are explored next, in Chapter 7, "Family Atmosphere and Family Values."

Exercises for Chapter 6.
Gender Guiding Lines, Role Models, and the Big Numbers
(Use in dyads, small groups, or class, or with clients.)

Gender Guiding Lines and *Role Models*

Draw a line vertically down a piece of paper. At the top of one column write "Masculine"; at the top of the other, write "Feminine." Now consider these questions: "What makes a man masculine/manly to you?" and, "What makes a woman feminine/womanly to you?"

Write down the qualities you associate to masculine and feminine under the two headings. Do this quickly, in an impressionistic way, so as to avoid socially dictated, stereotypical terms.

Next, fold the paper in half, crosswise, and under the appropriate Masculine and Feminine columns, write the words "Father" and "Mother." Under these two headings write your answers to the following questions concerning your pre-school and early grammar school years: "What kind of man was your father when you were a child, up to age nine or ten?" and, "What kind of woman was your mother when you were a child, up to age nine or ten?" In the absence of Mother and/or Father, you can still carry out this exercise. It is likely that, like most children, you filled in the blank with private notions about the missing parent. If not, and if there were parental substitutes who were important to you, either on the plus or minus side, write what of kind of men or women they were.

Search for matches. Unfold the paper. Compare the words used for "masculine/manly man" with the words for "Father," and the words used for "feminine/womanly woman" with the words used for "Mother." Explore the meanings of each word to identify as many words as genuinely match, as synonyms or as nuanced similarities.

Matching words represent *gender guiding lines*, that is, those ideas about the meaning of gender that reflect your basic convictions, ideas that seem to be destined and changeless, that, however, once uncovered and brought into awareness, can be altered or changed.

Search for differences. After a thorough discussion of possible matching words/thoughts, turn your attention to the words that do *not* match. These attributes and qualities represent *role models.* Ask yourself, "Who was there in my childhood world with these qualities? What adult did I know, or know about, who was like this?"

Discuss the role models and what there is about them that made them attractive (or unattractive) to you, with the awareness that these are persons you chose either to emulate because of their positive qualities, or to avoid emulating because of their dystonic qualities. *Role models* are chosen, not destined, so can be changed to more useful models if those chosen in early life are no longer desired or appropriate.

Consider the connections available in this exercise between youthful impressions of the meaning of gender and current convictions, evaluations, biases, and expectations for your sex and the other sex.

The Big Numbers

Reflect on major events or turning points in your parents' lives. Consider events in the life of your same-sex parent. Do you recognize any parallels in experiences at those same ages? Or, if the age marker is *prospective*, consider the character of your feelings of anticipation as you approach it. Also consider your expectations as your partner of the other sex approaches markers set by your parent of the other sex.

Chapter 7. FAMILY ATMOSPHERE AND FAMILY VALUES

Family atmosphere and family values are terms we use to describe qualities of the social environment (the field) in which the creative child worked out a style of living. Our goal is to awaken the client's awareness of how the present style (now called into question, and subject to reexamination) was once a solution to the problem of successful adaptation in a particular childhood situation. Here, as elsewhere in the psychoclarity process, we are striving for an appreciation of context that will facilitate understanding, believing that errors that are understood will no longer be seen as inevitable ("That's just the way I *am*!"), and can therefore be seen as no longer necessary.

By *family atmosphere* we mean the quality of emotional and affectional exchanges between and among family members, and the resulting quality of the social field in which the child develops a *style of personal interaction*.

By *family values* we mean matters of importance both to Mother and to Father (not, therefore, gender-limited in their application), with reference to which the child develops a *personal code of values*.

Family Atmosphere

Of all the terms we use in lifestyle assessment, family atmosphere belongs most clearly to the art of interpretation; it does not pretend to a place in any science of measurement. Following Dreikurs, we use this phrase to evoke a sense of the "climate" of the client's childhood household.

A meteorological metaphor can bring an old frame of reference into a fresh light and perspective, making it more accessible to understanding and reevaluation. Homely and vivid imagery will

212

be the most useful for this purpose. For example, everyone is familiar with what is meant by a "warm" greeting, a "cold" formality, or a "heated" argument. In the same way, everyone knows what to expect in "stormy" relationships, a "suffocating" situation, or a "climate" of friendliness.

The use of such terms to refer to the family atmosphere highlights the way family members were characteristically disposed to relate to one another. The relationship between Mother and Father (or parental substitutes) was, of course, primary in this. Other adults resident in the household either reinforced the pattern established by the parents' relationship, or offered examples of contrasting possibilities.

We rely upon picturesque language in lifestyle assessment because we need devices to reawaken the client's creative imagination. Bear in mind that we want the peculiarities of a style of living to be recognized as intrinsic to an adaptation originally worked out in a situation in which they worked well enough to seem appropriate. However narrow and unsuited for successful engagement with life in the wider community now, these can be acknowledged as having been effective in adapting to the climatic conditions of the client's childhood world.

A loud and quarrelsome family life, full of outbursts, insults, and tears, may have established the background for a style that is closed and cautious, as if "buttoned up" against the unpredictable fierceness of the weather. For another person, coming out of the same family (or another family with a similar atmosphere) the alternating scenes of conflict and brooding may have provided a training ground for theatricality, with rehearsals of intense feelings over a wide spectrum. Such histrionic styles were often worked out apart from clear thinking, perhaps even as a substitute for rational processes, since in such households systematic, deliberate, and consciously cooperative problem solving may be assumed to have been at a minimum. Rationality leads to agreements, and those who want victories are not interested in agreements.

A "calm" atmosphere may characterize a house in which attitudes and exchanges are pleasant, "balmy," and conducive to secure feelings and playfulness. In another house the use of this word to describe the atmosphere may refer to a "calm before a storm," full of electricity and apprehension. A child growing up in an atmosphere of the second kind may develop a suspicious but patient style of studied composure in any situation in which there is no evident agitation or disruption, as if to say, "There's nothing I can do until the storm breaks and I'm able to see which way the wind is going to blow." Another child may practice a surly, forbidding style, as if to say, "If we're in for a storm, watch my thunder!"

Remember that these meteorological metaphors are useful only when they are useful to the client, and when they fit the information presented by the client in the inquiries. Moreover, the fit should be comfortable, not forced. Just because analogies drawn from descriptions of the weather are often the most dramatic and powerful does not mean they are the only ones that can be used. Often there are more suitable ways to convey the characteristic feeling of a given household. For example, in *The Summary of the Family Constellation* in "The Case of Janice," reported in Chapter 11, we say, "Janice grew up . . . in a family in which the atmosphere was constraining, as if the parents were concerned to hold the energies of each of the children in check."

To identify the family atmosphere with a few key words or phrases, take into consideration what the client has told you about the relationship between the parents during the course of the lifestyle inquiry. The questions about other significant adults and the family milieu may also provide clues. (This information is elicited in Items 1-13, pp. 13-17, under "Special Diagnosis" in the *IPCW*.) For an example of the sources of family atmosphere information, see these items in "The Case of Dan," reported in Chapter 4. For examples of the way we include references to family atmosphere in the course of summaries of the family constellation, see Chapter 13. In the list below, also, we offer an assortment of words and phrases we have used in various cases:

[The client] grew up in an atmosphere that was:

"clouded over by Mother's fearfulness and sensitivity, and relieved by the occasional breakthrough of sunniness provided by Father's exuberance, gregariousness, and affection."

"close, warm, still, and at the same time, isolating, as in a hot-house."

"generally warm and conducive to the growth of self-confidence."

"tense and volcanic, though never threatening the possibility of breakup or loss, and therefore endurable."

"invigorating and warm, with each of the children feeling well cared for and able to develop without constraints."

"balanced between Mother's smothering warmth and Father's forbidding coolness."

"somber and discouraging, with Father playing the part of rainmaker with his squelches, and Mother maintaining a damp and covert rebellion against Father's rule."

"one of relentless striving, as if under a steely grey sky."

"turbulent, as if the family was situated on a storm front of conflicting currents of hot temper and icy disdain."

Family Values

At the beginning of this chapter we defined family values as having to do with matters of importance to both Mother and Father. It would be impossible to compile a complete list of such values as one might encounter them in work with clients. Even so, we can offer a partial list of those we commonly find in order to illustrate the variety of issues, each one important to someone, that parents sometimes strive to transmit to their children, and sometimes convey without striving, by those actions that speak so much more loudly than words. It must be noted further that family values are not necessarily the virtues commonly recognized as humane. Brutality, for example, can be held up as a good thing, as can theft and other deceitful maneuvers. In some families, such activities can be touted as values. A broad spectrum of behaviors, private in their understanding, may be opposed to the welfare of others, or even to the general welfare.

Education	Distrust outsiders
Religious observance	Hard work
Don't draw attention to yourself	Manners
Honesty	Don't complain
Loyalty to family	Better yourself
Order	Be the boss
Hospitality	Propriety
Deference to adults, teachers, "betters"	Serve others
Obedience	Fair play
Independence of thought and action	Do your duty
Do your best	Perfectionism
Be first and stay ahead	Don't take things seriously
Perseverance	Trust to luck
Thrift	Competence
Appearance	Self-assertion
Take what you can get	Belittle losers

Keep in mind that family values are defined as matters of importance to *both* parents. Otherwise, values held by either parent apart from the other should be considered as elements defining the gender guiding lines (Chapter 6).

The family values refer to standards that are held up before all the children, with the implication (or explicit statement) that, "These are the things that are important in this family, whatever others may think or do in other families."

This means that family values operate as imperatives. None of the children can ignore them; on the contrary, we see each child taking up a position toward them, either overtly or covertly, whether in support or in defiance. These are the issues that all the children in the family had to deal with in the formation of their lifestyles.

We identify these values in the lifestyle inquiry, in asking "What was important to Mother? What was important to Father?" under "Special Diagnosis" in the *IPCW*. Where values are alike for each parent, we assume we have identified a family value. Sometimes, however, the situation is not so clearly stated because clients themselves are unclear about the values of one or both of the parents. Then we may gain more information by seeing how the children of the family (where there are siblings) lined up with respect to various issues. For example, if in a family of three children two of them now hold graduate degrees and the third one dropped out of high school, we may assume that education was a family value, which two of the children accepted and deferred to, while the third child defied it.

The fact that a client reports what each of the children did in respect to any particular issue must of itself suggest to us that this was a matter of general importance, and therefore a family value. As an example, one client said, "George was president of his graduating class the same year Susan was editor of the yearbook. Marilyn was on the school council all through high school. When I came along, and just wanted to hang out with my friends, nobody knew what to do with me." We identified community engagement as a family value.

To sum up, we ask explicitly for information about what was important to each parent and, we ask not only for a description of the activities of each child and how each distinguished himself or herself from the others, but also the extent to which each of them accepted, rejected, or modified the family's standards and values. See the section of the *IPCW* concerning the children; these entries should be excellent resources for the therapist in helping the client toward understanding the family values and the position the client took with respect to them. Chapter 13, "The Family Constellation," provides further examples of the usefulness of this material.

Exercises for Chapter 7.
Family Atmosphere and Family Values
(Use in dyads, small groups, or class, or with clients.)

Family Atmosphere

1. Write a paragraph or two about how you perceived the family atmosphere in your childhood household before age nine or ten. Use and expand upon climate-related terms, for example, "The atmosphere in our house was cold, unwelcoming. I didn't bring my friends home" or, "Our house was usually sunny, but there were occasional storms — when that happened we scattered!" Describe the part played by each parent (or other caregivers in the absence of parents) in creating the atmosphere. What happened? Who was most affected?

2. Using colored pencils, draw a scene illustrating the family atmosphere. Explain the scene, describing what is going on, noting the part played by each person represented in the picture, especially the roles played by the parents (or other caregivers in the absence of the parents).

Family Values

1. Write a paragraph or two answering the question, "When you were a child, up to age nine or ten, what was important to both your parents (or other caregivers in the absence of the parents)?" Remember that values *not agreed upon by both parents* are not family values, but instead form aspects of the masculine or feminine guiding lines. Elaborate on the family values, for example, "Education was the most important value to Mom and Dad. This meant getting a college degree. They preached to us that without it, you can't get a good job. College was all about preparation for work."

2. Using colored pencils, draw a family crest in imitation of an heraldic shield. Consider what issues formed the family values in your childhood, up to age nine or ten. Divide the shield into sections for these values; for example, if your family emphasized camping, provide a section to represent camping with a tent or other object. Continue to form sections for other values. At the top of bottom or your crest, write a *motto* that reflects the central theme emphasized in your family (for example, "Be prepared!").

Conclusion

Children are born into a world that is already inhabited, and into families that already have histories. In these families they encounter standards for a quality of interaction with others in the family atmosphere, and standards for distinguishing between what is important and what is not in the family values. This is the basis for the folk wisdom that says, "The fruit doesn't fall too far from the tree." Even so, we look to find the individual variant, and to see how the child's creative power made use of these givens.

In the next chapter we move from considering the person's place as studied against the background of the entire family system to give our attention to the subsystem of relationships among the siblings. Our task there is to understand the various meanings children attached to the birthorder positions they held in their families, and the various attitudes they adopted with respect to those meanings.

Chapter 8. PSYCHOLOGICAL BIRTHORDER VANTAGE

In this chapter we will consider psychological vantages and situational variables relating to birthorder as set forth in Figure 7, page 221.

We use the term vantage instead of position (more common in discussions of birthorder) to emphasize the distinction between psychological birthorder and ordinal position. We want to consider how each child's situation in the family provides a unique perspective from which to view self, others, and the world. Thinking in terms of perspective rather than position helps the interpreter understand the particular use this one person made of a particular place in a particular family.

This terminology also helps to avoid an implication of causal determinism in birthorder. "You are bound to do [whatever]; you're a firstborn," is an example of typological errors that obscure the genius of Individual Psychology, "the task of [which] is to comprehend the individual variant," as Adler said (p. 180). We pursue this task by the original Gestalt method of locating figure against ground. When the discipline of this method is ignored in simplistic efforts to fix cause and effect, the significance of birthorder in understanding a person is obscured, and dismissed by serious investigators as on a level with astrology or numerology.

To make use of the information provided by an exploration of birthorder, we must see it as indicating *probabilities* to be tested against the particularities of a given family constellation. These probabilities then contribute to a frame of reference for interpretive guesswork that is both useful and encouraging.

PSYCHOLOGICAL BIRTHORDER VANTAGES
AND SITUATIONAL VARIABLES

The value to the psychologist of considering birthorder is that this feature reveals the primary and original vantage from which one perceives and evaluates self, others, and the world, and in which the child forms convictions about what is required of him or her to move toward *success,* as personally defined in the child's subjective sense of her or his total situation. Vantages and variables listed here are meant to be suggestive, to provide a framework for inquiry, and to emphasize that *no one is to be regarded as a type.*

Psychological Vantages to Consider

1. firstborn only

2. firstborn

3. secondborn

4. middle

5. thirdborn

6. youngest

7. baby

8. laterborn

Situational Variables to Consider

1. parental birthorder vantages

2. age differences, and (in a large family) how the children grouped themselves

3. sex differences, including the child's impression of the value placed on the two sexes by the parents

4. children's health

5. defaulting siblings

6. death of a sibling during client's childhood, or before client's birth

7. mother's unrealized pregnancies

8. introduction of step-siblings

9. parental favoritism

10. child's perceptions of parents' marriage "arranged" by a pregnancy

Figure 7.

Situational Variables

A complete understanding of birthorder takes the following variables into account: parental birthorder vantages; age differences, and, in a large family, how children grouped themselves; sex differences, including the child's impression of the value placed on the two sexes by the parents; children's health; defaulting siblings (whether the child failed to meet parental expectations, was confined to an institution, or otherwise defaulted); death of a sibling during client's childhood, or before client's birth; mother's unrealized pregnancies, whether through miscarriage, medical abortion, or stillbirth; introduction of step-siblings; parental favoritism; child's perceptions of parents' marriage "arranged" by a pregnancy. In anyone's particular experience of childhood there may be factors not included in this list that must also be noticed and attended to.

Parental birthorder vantages. Part of a child's interpretation of self, others, and the world has to do with parental attitudes toward children born into the same birthorder position as one of the parents. According to pediatrician Peter A. Gorski (as cited in Klass, 2009), "Too many parents are haunted by experiences both good and bad that they identify with their birthorder" which can lead to "self-fulfilling prophecies" as to how a child in the same position will function now and in adult life. A parent may identify with the child, projecting onto him or her expectations arising out of the parent's own experience in childhood. If the parent managed the position successfully, he or she may expect the child to achieve success as well, and be flummoxed, upset, angry, even rejecting, when observing the child's missteps. Similarly, if, as a child, the parent saw others from a self-perceived position of disadvantage, the parent may feel sorry for and pamper the child in an effort to make it up to the child for what the parent believes he or she is expected to suffer.

Age differences and, in a large family, how the children grouped themselves. Age differences always matter, but matter especially if there is a gap of six or seven or more years between siblings. In such a case, the older child, now well-established as a firstborn only child, and having entered the social world of the school and neighborhood, is not necessarily hampered by the birth of a sibling; indeed, the older child may enjoy having another child in the family to relieve the pressure of parental attention.

The situation differs when a second child is born within a narrower timeframe. A second child, born close on the heels of the firstborn, may strive to unseat the pre-eminence of the older sibling in what Adler (1964) characterized as the *dethronement* of the firstborn (p. 377). Such an exchange of places is usually experienced negatively by both of them. Although the older may be relieved to be out of the spotlight of leadership, and the younger may be pleased to be the pacesetter, the older most often feels defeated and demoralized by the displacement, and the younger may resent having to take the risks of the untested lead.

In large families there may be several groups of children. Most clients know at once what you mean if you ask, "How did the children group themselves?" or, "How many groups were there among the children?" Here is how Grace, the third of five, answered:

Tim and Phil were a pair. They were the big boys. Then I came along, followed by Maureen and Edgar, the little kids.

Grace may mean that she took her position as first of the three "little kids"; or, a position as the older of the two girls, with Edgar in a class by himself as the baby boy; or as middle child between the older boys and the "little kids." There may have been other possibilities open to her as well; only Grace could tell us the position she claimed as her own. The importance of the question is to establish the client's vantage, as the client saw it.

In the case of twins or other multiple births, children sort out birthorder. It cannot be designated; it must be claimed. Parents will say, "The boys don't know who was born first," but, of course, the boys know — in part because they are introduced as "our boys, Fred and Charlie."

Sex differences, including the child's impressions of the value placed on the two sexes by the parents. Children are alert to any discrepancy of value based on gender. In some families children of both sexes know themselves to be valued equally by the parents; in other families one sex has greater value than the other. These attitudes affect the ways parents interact with their children, and the ways in which the children interact with one another. The cultural heritage of pyramidal forms of social organization based on conquest and subjugation persists: Sons are still often assigned a higher value than daughters. Disturbances experienced in the souls of children when they confront the challenges of the emerging democratic era, its demands for cooperation, and its insistence on equal respect, can be profound and destructive. Girls do not move cheerfully toward positions of submission and subordination; boys do not move courageously when they believe they must achieve and sustain positions of domination.

In a time of social change, when autocratic forms are challenged by the democratic critique, but when new democratic forms have not yet been established clearly and firmly in their stead, there are many variations in the experiences of inequality among children. Here is one male client's report:

> My sister was a princess. Everyone doted on her and I felt ignored. Men in our family weren't worth much, even to themselves. Women could do no wrong. I decided to make the most of a bad position: I was a little monster and they couldn't do anything with me!

In some traditional families the genders are sharply distinguished in role expectations, but an attitude of respect for the equal dignity of each is maintained. In a family of this kind there can be both a firstborn son and a firstborn daughter, each of whom exercises a role of leadership in a separate domain; however, "Separate, but equal" is a doctrine increasingly difficult to maintain. Here is a female client's report:

> There were the boys, and there were the girls. We each did our own things. We were a large family; we even had separate sittings at supper. The boys were fed first, of course, and then the girls. I hated that.

While most children sort sexual matters out successfully, especially in highly gendered families some may have a private conundrum, un-understood at the time, "If it's not acceptable to be who I am in my sex, then what am I, and how am I to operate?" These children may continue to struggle with this question, especially as they move into adolescence and adulthood. They may move away from their native sex and gender altogether, toward other sex and gender possibilities; some may go through a lifelong process of questioning their gender identity, orienting and re-orienting themselves (Vitello, 2006; Warsett, 2010). Whatever the later outcome, in childhood they are of concern to their siblings, who may have feelings of sympathy and support for them, especially if a sibling is subject to bullying (Turnbull, 2010), or may feel pity for them, or may look down on them, or may choose to ignore them altogether, cutting them out of everyday childhood activities.

Children's health. A sick or invalid child affects birthorder vantage, altering the field in which the children are moving to make their places. Even a mildly ill child can train for (and be trained by the others for) the position of invalid. One of the siblings may seize the chance to play "little mother" to this child, thereby ensuring a place of importance both with the sickly child and with the parents. It is then in the little mother's interest for the sickly child to stay sick, with the caretaker training the one cared for to continue in the role of invalid. Another child in this family may see an opening to the place of being his father's "tough little guy," as if to make it up to his father for his father's distress over the sickly one.

Defaulting siblings. To default means to fail or to avoid an obligation. Children born into any ordinal position in the family may evaluate that position as psychologically too demanding, as "too much for me," yielding the place to another sibling by default. Children may default through institutionalization, developmental disability, disqualifying misbehavior, or even general ineptitude.

In adolescence, excitement seeking in alcohol, drugs, dangerous driving, and/or harmful sexual activity may lead to a young person's defaulting from an expected place in the family (Walton, 1980).

The defaulting of a twin can be particularly disturbing. Composer Allen Shawn (2010) was separated from his severely autistic twin sister, Mary, when she was institutionalized at age eight. He writes:

> [Her disappearance] left a kind of ocean of disquiet in me that manifested itself in panic attacks and a lifelong struggle with agoraphobia. . . . As her twin, it was doubly hard for me to know how and where to draw the boundary between her nature and mine, between the inherent strangeness of being a person and the kind of strangeness [Mary's] that I saw as banishment from normal human society. . . . It wasn't until I reached late middle age that I could even begin to acknowledge that being Mary's twin was a central fact, perhaps *the* central fact, of my life. (p. 8)

Death of a sibling, during client's childhood, or before client's birth. The death of a sibling can be regarded as a form of defaulting, responded to variously by the others, from loss and mourning to anger and resentment; whatever the case, when illness overtakes a healthy child who then dies (or is developmentally disabled), the constellation may change and each child's vantage may be altered. Even a sibling who died before a child was born can play a part in that child's imagination. One woman described her childhood image of her older sister, whom she had never seen: A neat and clean girl, with spotless white shoes and stockings, sitting on one of a pair of straight chairs on a cloud, next to her grandmother, who sat on the other chair. This picture of "my sister, who art in heaven" was onerous to the living child, who could never stay as clean and neat and properly seated, or so free from the danger of scolding or criticism.

When a healthy child is born following the death of a sibling, parents may stand back in wonder at their miracle baby, or they may hesitate to accept the new healthy child, withholding love and affection out of fear that this child, too, might die.

Shelley Burtt (2010) was such a parent. She lost a child to illness at a little more than two years old. Later, after many months spent in the process of adopting, she and her husband took their new baby home. But, without understanding why, Shelley remained distant from this much-wanted "replacement" child. In a reminiscence of that period, she writes:

> I still ache when I remember how at each milestone, this little boy was held up to his absent brother — the first step, the first tooth, the first word. I didn't realize, until 27 months and 14 days came and went, that I was counting the days till the second son would outlive the first . . . [from then] Ryan made his own history, a gift I should have given him from the first but could not. [See the Big Numbers, Chapter 6.]

Mother's unrealized pregnancies. A child's training and self-training is affected by an awareness of other, unrealized pregnancies. Like a healthy child born after the death of a sibling, the parents may view the child whose birth was preceded by a series of miscarriages or stillbirths, as "God's gift." Such a child may grow up feeling invincible, with special immunity to trouble.

Other possibilities are that the parents treat the living child as so precious that it must be protected from any danger, as if the least difficulty could be disastrous. This child's convictions, deformed by limits placed upon experience, may be, "Life is dangerous; therefore, I need the protection of others." Such a child may also be especially pampered, and believe that the world should be in his or her service.

Introduction of step-siblings. Established family patterns necessarily undergo changes when step-siblings enter the group. Age and sex differences must be taken into account in any troubles that may ensue. Younger children may have the advantage of flexibility in the new situation, depending on the extent to which they have rehearsed a particular birthorder vantage. Older children may not be as resilient as younger ones, but, since they have activities outside the home, have options not available to younger ones. The factor of parental favoritism may play a role in the success, or difficulty, or failure, of the blended family.

Parental favoritism. Though parents protest they have no favorites, children know they do and who the favorites are. Favoritism may arise, for example, when parent and child share a particular talent or enthusiasm (music, sports, art, shop); or, when one of the children resembles a loved father or mother of one of the parents; or, when a child is born into the same birthorder of a parent; or, clings to one of the parents in a way that flatters the parent, as if the parent were indispensable to the child.

Rehearsed preferences may shift with the entry of step-siblings, causing tension and problems in the new family. For example, a child may be displaced from the position of the favorite if a more energetic, determined, attractive child enters the group, especially if the child courts the parent and the step-parent encourages the child in these directions.

Children's perception of parents' marriage "arranged" by a pregnancy. "Who cares?" might be the reaction to this subject upon first encountering it — in contemporary culture, it hardly seems worth mentioning. But a child who brought about such a marriage, having been told about it by parents, relatives, or neighbors, might think of it as an explanatory principle for anything that goes wrong. Unfortunately, parents can and do tell their children, "If it weren't for you, I wouldn't be here! I'd be doing something *worth* doing!" Or, "I could have been _____, if it hadn't been for you!" Children hearing such exclamations may come to believe that simply by being born they wrecked the parent's life, and carry this shame into adulthood.

Children respond variously in other situations. A child might announce guilelessly, "If it weren't for Susie, Mom would have married someone *nice*," or, another child, knowing the parents "had" to marry, might announce joyously, "You know what? Buddy made Mom and Dad get married!"

With these special considerations in mind, we may, without too much remaining danger of mistaking them for determinants, review some of the probabilities connected with the various birthorder vantages.

The Vantage of the Firstborn and Only

Firstborns who never have siblings are never dethroned. To be spared the presence of (and possible competition with) siblings is also to be unfamiliar with the skills required to dicker and negotiate with them in the give and take, rough and tumble of childhood politics. Their major common complaint about their childhood situations has to do with loneliness. Even those who deny any consciousness of this loneliness often reveal it in their remembrances of wanting impatiently to be grown up, and so, to be like the other members of their families. Another revelation of it is in their commonly stated ambition as adults to have more than one child.

The advantages of being at the center of the parents' worlds belong to the only child. Never to have to yield any part of that, or to share it with other children, can be a source of confidence.

However, this central place can also have the disadvantage of carrying with it a sense of burden. "The hopes and fears of all the years" of their parents' ambitions, difficulties, and decline are laid on them. They know that if they fail, the parents have no other children to compensate. It is consequently not surprising to find a consistent theme in their styles expressed in the thought, "It's all up to me." This idea is experienced with increasing uneasiness as their parents move into old age. They have no sibling to confer with over what to do about "Mom's failing memory," or "Dad's insistence that he can drive, when everyone knows that he can barely see." Because they often exaggerate the meaning of the common human sense of uniqueness, regarding it as a kind of impenetrable privacy, they find it difficult to discuss these and other worries even with a devoted spouse. They too easily assume, "I must be the only one who thinks and feels this way."

During childhood the exaggerated sense of being different can be painful to them in dealings with other children. Having spent their early years with adults, and therefore apart from peer level problem solving, they often entered the world of the neighborhood and school without training for getting along by going along, and may have felt lost among the others. The bright side of this experience is that they understood adults better than the others did, and so were better able to fit into the organizational regimen of the school. They were better able to figure out what the adults wanted, and were also better able, or at least more readily disposed, to deliver it. The dark side is a frequently reported sense of having been a misfit or "oddball," of always having to guess about what was going on among peers and, when the guesses were wrong, of being embarrassed by their *faux pas*.

Many who grew up as only children have come to expect and resent a common assumption of others that they must have been spoiled by their parents. In fact, the pampered only child is a rarity in our clinical experience. Those who grew up sharing the world with siblings may believe that their only-child chums were spoiled because they saw them as having more things of their own. The objects of their envy see things differently, however. Here is a report from a client, Grant:

Yes, I was the first one and sometimes the only one in the neighborhood who had a computer, or the latest electronic gadget, and so forth. And, yes, I had a room of my own; in fact, I had the whole third floor, complete with bathroom, all to myself! BUT, the key words are "all to myself." What good were those things to me when I had no one to play with? I got furious when people told me I was spoiled. I *always* did what I was told, studied hard, took care of myself, and was a good kid. Is that spoiled?

An only child can have a place in the triumvirate with Father and Mother, but only as the one at the awkward disadvantage of not knowing as much as the others. In this trio, the only child may experience a pseudo-sibling relationship with the same-sex parent, as if mother and daughter were the two girls in Father's life, or father and son the two boys in Mother's life. Such a child may train and be trained in secondborn operations and expectations to some extent.

Encouraged only children are characteristically self-reliant and competent. They enjoy working alone in solitary problem solving in which they mull things over for long periods and then announce solutions. They are people who had to figure things out alone in the course of growing up, without much consultation or assistance; necessity was mother to the invention of a style whereby this might be done.

The Vantage of the Firstborn

The perspective common to these children is upward, forward, toward the adult world and its standards, and away from the world of the other children, except when sharing the concerns of the adults over them.

Remember that a firstborn was, for a time, an only child. Firstborns often retain some of the characteristics of onliness, including loneliness. Such persons have a sometimes poignant sense of personal uniqueness. They may practice self-reliance as if sensing it to be necessary, as people do who are accustomed to being left alone to solve problems. They may be less at ease with members of their own age cohort than younger children are, sometimes feeling prepared only for places among the elders of any group to which they belong.

Adler noted (p. 378) that firstborns tend to be conservative in outlook, because they are nostalgic about the lost golden age of their only child status, from which they were dethroned by the birth of a younger sibling. Adler thought that this fictional overvaluing of the past brought with it a greater respect for tradition, concern for law and order, and caution in assessing innovations. It is true that most firstborns experienced themselves as very important persons in their "good old days." Indeed, they often were. They made wives and husbands into mothers and fathers, and (when they were also the first grandchildren) promoted the parents of their parents into grandparents, another important social role. More photographs were taken of them than were taken of their younger siblings, and their families kept more complete records of their first steps, first words, and first teeth.

Laterborn children often complain about the special status of firstborns, unaware as they are of the negative features of firstborn experience. A great deal can be expected of firstborn children. For example, a younger sister was called The Baby at a much later age than that at which her firstborn sister had become The Big Girl. The title, "Big Girl," came with "all the rights and privileges thereunto appertaining," but with some matching burdens as well. Many firstborns become perfectionists in the effort to meet the challenges of parental expectation (and later, to do what they believe is required of them if they are to have a place among others). They are often constricted by righteousness, unable to see how they are of any value unless they are doing something wonderful or — even more important— admirable.

The sense of obligation to do only things that are outstanding may lead to hesitation, stalling, and a dread of engaging in anything. After all, whatever anyone does can always be improved upon. A style of procrastination (sometimes disguised as preparation) may burden firstborns. They may experience anxiety (an expectation of defeat or humiliation) in the presence of opportunities they first define as obligations and then feel unable to refuse.   Achievement, leadership, an interest in exploration, and the firstborn vantage all go together when these children are encouraged, appreciated for what they can do, and respected for how they choose to do it, without the burdens of unrealistic expectations.

Adler was a secondborn, and described firstborns with more than a hint of the asperity, impatience, and failure to sympathize that many secondborns feel toward their older siblings.   His observations about firstborn characteristics should therefore be read with some care.   All of Adler's comments on the various sibling positions are, in fact, suspect as general rules; they should be studied only to show how he framed his hypotheses for reaching an understanding of the unique variant, and to sharpen awareness of the differences in phenomenological field made likely by different birthorder vantages.   The study of Individual Psychology can contribute greatly to clinical sensitivity; it is not a substitute for it.

The Vantage of the Secondborn

Central in the awareness of the secondborn child is the presence of another child. It is as if the secondborn knows from the beginning that, "someone more or less like me is here, a pacesetter scouting out the territory of life, discovering what it holds and what it requires." The secondborn hears the firstborn using the language, sees the firstborn relating to Father and Mother, and watches the firstborn go out into the neighborhood and the school, as if in an exploration of territories beyond immediate reach. Before they feel ready to meet the challenges themselves, secondborns are listening to the firstborns' reports on how it is "out there" in a world for which they are still (often impatiently) preparing.

If the firstborn is of the other sex, the second may take the vantage of another first, the eldest of my sex. Then, if the children see a friendly relationship between the parents, we can expect them also to be friendly (or at least polite) to each other, to avoid direct competition, and to find separate and distinct spheres of successful activities.

When the secondborn is of the same sex as the first, the children are more likely to make other choices. Number one may regard number two as an interloper, and attempt to thwart the challenger's advance. One way to achieve this is to keep number two in the position of baby, either by belittling the younger child, ordering it about, or (more subtly) conspiring to pamper, protect, and serve the younger so that the secondborn comes to feel dependent upon the first.

Success is not guaranteed to any of these strategies. Nothing a firstborn does to meet the challenge of rivalry can avoid the possibility of spurring the rival onward. Held back, secondborns struggle all the harder to advance; ruled, they are all the more likely to rebel.

Even to say this much, however, is to risk obscuring a sense of the variety of ways that can be taken by a child, perhaps especially a secondborn child. If firstborns are more concerned about the world of adult rules and standards, and if secondborns are at pains to distinguish themselves from the firsts, we can expect a greater degree of openly displayed creativity and spontaneity in their styles. Conformity and innovation do not mix easily in the same person. To gain the attention and enlist the active interest of a firstborn we say, "We don't find anyone who is able to do this task in accordance with the specifications." To a secondborn, the challenge lies in hearing, "Until now, we have found only one way to do this satisfactorily."

The Vantage of the Middle Child

While the advent of a third child alters the objective position of the secondborn to that of middle child, the secondborn may hold fast to the vantage of the secondborn child and never become a psychological middle child. When the second does shift to the vantage of the middle, however, a new range of possibilities presents itself.

If, until now, numbers one and two have been engaged in a struggle for preeminence, the newly deposed and hitherto baby of the family may feel forced to fight a war on two fronts, alternately contending with the firstborn for the honors of leadership, and with the new baby for the exemptions and protections of littleness.

Or, if the parents are pressuring the older child to excel, and fretfully fussing over the younger one, the middle child may be aware of being the only one free of the burdens of parental management.

In either case these children often feel squeezed, and have to evaluate the feeling. An active child may feel squeezed into an awareness of inequity, forming a guiding ambition to shape the choice of a vocation to contend against injustice. A less active counterpart may see the need to struggle for justice as in itself unfair, and adopt the posture of a victim. Sometimes the squeeze feels like buoyancy, pushing the child up and out to a greater sense of freedom. It may also be felt as pressing the child down and out toward the threat of insignificance, against which the person fights throughout life. The squeeze can also be experienced as buttressing, with the middle child reporting enjoyment in having a supporting role played on either side, and shining as the star of the show, or "the rose between two thorns."

The variety of these possibilities probably helps to account for the middle child's so frequently exhibiting an exquisite capacity for empathy. These children know how it feels to be under the heel of someone ahead of them, and what it is like to have someone of less experience relying on them for care and guidance. They know what it is to look up to others, what it means to look down on them, and how it feels to be on the receiving end either way.

In subtle ways at first, and in more and more practiced ways as they grow older and the other family members are separated and scattered, the middleborns tend to operate as the central switchboards of their families, through whom the others stay in touch. Remembering birthdays and anniversaries, they remind those who (counting on the middle child to remind them) are likely to forget. When there is an estrangement in the family, or where delicacy may prevent one from speaking to another, this person may receive a call and respond to the plea to negotiate a settlement or to convey a difficult message.

The Vantage of the Thirdborn

"The enemy of my enemy is my friend," according to the adage. The alliances typical between thirdborn and firstborn children tend to bear this out. Siblings can be expected to compete to some extent, but the more intense forms of competition are most often between firstborns and secondborns. In the experience of many thirdborns, consequently, this conflict was there from the beginning. One result of this derives from the fact, noted above, that the middle child is not only fighting against being held back, but also against relinquishing the advantages of baby. First and third may come to see something attractive in each other, having as they do a common enemy. In the case of the third, admiration of the first is demonstrated by imitation ("the sincerest form of flattery"), understandably adding to the first's enjoyment of the third.

A second consequence follows out of the apparent endlessness of the fighting between first and second. With no sign that either of them is going to emerge as winner, neither appears to be gaining anything from the struggle. "Therefore," many sensible thirdborns conclude, "there must be a better way." They may decide that nothing is worth fighting about, that charm is more effective than force, and that it is wiser to beg forgiveness after doing what you wanted to do than to ask permission beforehand and risk refusal or opposition.

Feedback confirms the wisdom of the choices: (a) The firstborn accepts the admiring third as a protégé, as if to say, "I got off on the wrong foot with the secondborn. I'm going to get it right this time, and stay ahead by being the mentor who teaches this one how to get ahead." (b) Charming is disarming, and the parents, already relaxed by it, relax further to see the little one being so well tutored.

The Vantage of the Baby

"Babies are made, not born," said Rudolf Dreikurs (personal communication, n.d.). Any child is vulnerable to babying. A firstborn, strongly favored, not necessarily by both parents, may be pampered, and demand and get special service. This is also true for any other child in the family, especially one for whom parents and siblings feel sorry, for any reason. Such a child can exploit their pity, as if challenging them to "make it up to me for what I suffer."

If this is a posture more readily assumed by younger children, it is because it is facilitated by co-conspirators among the older ones, who may imagine it to be to their advantage to shut a baby out of competition by keeping it babied. This is done by translations of baby talk; constant fussing and care; overprotection against the possibility of any hurt; and, most effectively, by providing services that make it unnecessary for a child to explore and learn, so thwarting its development. Add a stream of warnings and cautions, degrading in their reminders of special status ("You're not big enough, smart enough, strong enough, responsible enough, etc."), and the babying may be complete.

Many children who trained and were trained to be babies exploit the position throughout their adult lives, as if to say, "You wanted a baby? You got a baby!" As families grow smaller there are fewer cases of the bachelor uncle or maiden aunt remaining at home. There are still many examples of the relative who never learned to drive and has to have a ride to any event, or who cannot hold a job, or who "goes to pieces" when there is any hint of being left alone to handle responsibilities or of being counted upon by others.

The Vantage of the Youngest

The baby is not always the youngest child in the family, and the youngest child will not always remain a baby. Where children are encouraged and their individuality is respected, the experience of the youngest child can be rich in opportunity. Older siblings may offer themselves as coaches of the youngest's efforts, or as champions of the child's aspirations. The youngest can also study each of the older children for examples of successful and/or unsuccessful strategies, finding a variety of useful role models, both positive and negative. The unique privilege they enjoy among their siblings of never being dethroned is a further source of confidence for these children. Never dividing their attention by looking backward to see who may be overtaking them, they are free to focus all their energies on what lies ahead.

It is not for nothing that so many sacred stories, legends, and folk tales celebrate the triumphs of the youngest child, as illustrated by the Biblical story of Joseph and his brothers. (Granted, Benjamin was younger than Joseph, but only in birthorder position; in psychological vantage he was an only child.)

Often encouraged toward success, the youngest may also feel spurred to it. There is some obstacle to dignity in a situation in which older siblings often refer to matters of which one has no knowledge and demonstrate competencies beyond one's present ability. More common, and more onerous, is the readiness of the older children to regard one's contributions with a condescending delight hard to distinguish from amusement. A youngest child may feel the need to struggle to be taken seriously.

Meredith, the youngest of eight accomplished children, reported the reaction of her siblings to the news that she had earned her doctor's degree. "It was funny," she said. "They were very nice, and seemed sincere in their congratulations. But I kept picking up something in their tone that said, 'Look who got a Ph.D.! Isn't that cute?'"

The Vantage of Laterborns

In a large family, laterborn children form their own groups, including a first; second or middle; third and youngest, or baby. A fourth child can take part as baby; or (paired with number three) as the younger of two; or as ally of number two (especially where number one has appropriated number three); or as "odd one out" between an older group of three and a younger group (in which case he or she may develop some characteristics of an only child). Innumerable other combinations are also possible, depending on age, sex, and other factors.

Case Examples

The following brief sketches illustrate some of the ways children found vantages for themselves among the others in their families. Additional cases are cited in Chapter 13.

<u>Firstborns</u>

Mary is a firstborn daughter, with a brother two years older. She was Father's favorite, and enhanced her special place as the only girl by satisfying Mother's demands as to household chores, homework, and practicing the piano; and by satisfying Father's wishes that she be happy with him and enjoy him. Mary's brother was ahead of her only in that he was older.

Paul is a firstborn son with a sister five years younger, and two brothers, six and seven years younger than he. He was in a class by himself, with the other three children in a class of their own. As the first, and the only child for five years, Paul had ample time to establish his kingdom. None of the other children threatened his position. He kept his distance from them, and saw to it that they kept their distance from him by his solitary habits of reading and tinkering, and by playing the part of "the one who knows all."

Sam is the firstborn of six. He stayed ahead of his brothers and sisters by setting an example of rectitude and restraint, never failing to do what was expected of him. As the family increased in size, Sam turned his attention to the wider world of the neighborhood and school, where he took a leadership position among his peers.

Lucille is the first of two sisters. She made a place with Mother by being "little mother," thereby also insuring that her sister remained The Baby.

Secondborns

Bill is the second of two boys, psychologically a firstborn and only child, owing to his brother's congenital disability. Bill experienced himself as one who could do no wrong in the eyes of his parents. They were delighted with their healthy boy, providing him with every advantage and encouragement.

Martha is the secondborn of twins, whose brother, Craig, is eight minutes older. Psychologically she is the baby girl, with an older brother who took the position of the firstborn son. Their parents had been told they would not be able to have children. They were therefore thrilled to have "one of each." They counted on their son for achievement, and they spoiled Martha, giving her whatever she wanted. Growing up as the center of attention, Martha didn't have to do anything except "be." She didn't make a place: She was handed a place.

Al is the secondborn, whose older sister was "Miss Goodie-Goodie Two Shoes." He found a place in the family by being the naughty boy, given first to teasing, jokes, tricks, and being cute; and later in the neighborhood and the school, to an escalation of these "darling" games into troublemaking of all kinds. Mother and Father didn't know what to do with him, took pride in his being "all boy," and hesitated to "squelch" his mischievous inventiveness.

Shirley is the second of two girls. She distinguished herself from her older sister by developing her artistic capacities, followed the family values in education and religion, and found a place among her peers by practicing ways of offering them amusement and entertainment.

Middle Children

Florence is the secondborn child, with an older brother and a younger sister, psychologically the middle child of the family. She took advantage of the middle position by being the family peacemaker and negotiator. In the neighborhood, and later in the school, she turned this ability to further good effect by making many friends and by involving herself in school politics.

David is the secondborn, psychologically the middle child, with an older and a younger brother. He felt squeezed by the other boys, who were close to him in age. Jake, two years older, was the achiever in the family; Ken, two years younger and the last of the children, was Jake's protégé and the family's "junior achiever." David made his place by being the most active of the boys and a daredevil in the neighborhood and school. He was outstanding in sports, being well-coordinated and able to run fast, exceeding the accomplishments of his brothers in this area. He further distinguished himself by doing poorly in academics.

Thirdborns

Karen is the thirdborn in a family in which a stillbirth and two miscarriages preceded her arrival. Her brother is seven years older, and her sister six years older. When Karen was five years old Mother gave birth to twin girls. Of the five surviving children, Karen is the sickly one. Mother had overprotected her, believing that she might not survive a severe illness in her first year of life. Karen had taken a place as the homebody, involved in Mother's activities, and now shared with her in caring for the twins after their arrival. *Psychologically Karen is a middle child*, between the older two children and the twins.

Cliff is the thirdborn child, with a sister four years older and a brother three years older, each of whom took firstborn positions. Cliff did not, however, take the psychological vantage of the secondborn of the two boys; on the contrary, he made his place in the family by being the non-competitive, charming, and friendly thirdborn child. While he wouldn't let anyone push him around, he was not aggressive. He was popular with all the family members, counted on to keep the peace and to keep his word.

Peggy is the thirdborn of three girls, psychologically the baby of the family. The others fussed over her and taught her to be cute. She was Father's favorite and, since Father ruled, Peggy was the Princess. The other girls mimicked Father's attitudes and behavior, and also pampered Peggy, as a way of retaining their own alliances with Father and inducing Peggy to accept an inconsequential position.

Conclusion

A birthorder vantage includes ad-vantages and dis-advantages. There are no "good" or "bad" positions that in themselves could have made us happy or unhappy in the situations into which we were born. In our ordinal positions we created the vantages from which we considered the world; our birthorder was part of what we saw from our vantage.

Awareness of psychological vantage established by evaluation of birthorder offers counselor and client opportunity to reconsider how training and self-training rehearsed an attitude from that vantage. We learned how to move among others in our original families, where we formed our basic impressions of how we fit in, how we stand out, and what is required of us to do either. It can be liberating to reflect on what we made of our positions there; to imagine that our birthorder positions were determinant, that they *made* us as we are, is to miss the point.

In the next chapter, we examine heredity and environment in an investigation of the sense children make of their other possibilities and opportunities.

Exercise for Chapter 8.
Psychological Birthorder Vantage
(Use in dyads, small groups, or class, or with clients.)

After reviewing the *vantages* and *variables* pertaining to birthorder (Figure 7, p. 221), and the discussion in the chapter, state your perception of your own psychological birthorder vantage.

Chapter 9. GENETIC POSSIBILITY AND ENVIRONMENTAL OPPORTUNITY

In *A Religious History of the American People*, Sydney E. Ahlstrom (1972) reports on a quarrel about the idea of original sin that, from 1833 until 1847, divided the congregation of North Church in Hartford, Connecticut. Ahlstrom cites (pp. 610-611) the minister of North Church, Horace Bushnell, as memorializing this dispute under the name of the "Nature-Nurture" controversy. Like other theological arguments of its kind this one was not so much settled as forgotten. The original antagonists died, and the issue that had been at stake between them seemed to die as well.

It should be a matter of some interest therefore to see that this debate, which took place in a local nineteenth-century New England church, has (at least in name) for some reason survived or been revived. Contenders in the new Nature-Nurture controversy appeal to the majesty of "Science" instead of Scripture, but with the same kind of confidence that they enjoy the support of an absolute authority.

Inspired by the provocative and original studies done by Nikolaas Tinbergen and Konrad Lorenz on fixed action patterns in animal behavior, the Nature partisans in this new form of the old quarrel have been further stimulated by the work of Edmund O. Wilson, Robin Fox, Lionel Tiger, and others in the vanguard of what is called sociobiology. As the name implies, this involves extrapolations from animal studies in speculations on how the forms of human social customs may have been fixed during our phylogenetic history.

In the current round of the debate, the Naturists hold that genetic determinants shape the forms of culture; the Nurturists answer that the operations of culture are sovereign in molding genetic material. By extension to personality theory, one extreme argues that everything regarded as distinctive of persons was encoded in their genes; the other extreme retorts that a person is nothing other than a learned pattern of observable interaction with the world.

There is clearly an irresistible charm to this argument. However, since it pits the proponents of two sets of equally unverifiable speculations against each other in an intractable way, and since the matter is never settled (and the argument is never dropped), we are entitled to suspect that something unstated is reflected in the two positions. Dreikurs maintained that the hidden debate was a contest between two radically opposed political visions of the ideal social order, neither of which is sympathetic to democracy: On the one side are latter-day supporters of the old aristocratic ideal, summed up in the phrase, "blood will tell." On the other are proponents of one of the utopian egalitarian ideals, with a model of the person as a *tabula rasa* of infinite possibility, infinitely malleable.

From the standpoint of Individual Psychology the argument rests upon a fundamental misunderstanding, and is therefore meaningless. So are efforts to work out positions between its polar extremes. For example, "Heredity *times* environment *equals* personality," one university professor of human development put it to one of the present authors. He might just as well have said, "None of us knows how to account for the uniqueness of personality." Adler (1963) distinguished Individual Psychology from any such efforts:

> We proceed quite differently from other psychologists who would like to apply almost mathematical values and who, when their calculations do not come out, try to find causes in heredity, a dark area which can accommodate almost anything. (p. 1)

In fact nothing can be seen to *form* personalities; on the contrary, the very term, personality, refers to someone who is *giving form* to experience. We want to examine an individual's style of movement in the use of genetic possibility and environmental opportunity.

A lifestyle rests on opinions, on evaluations of a genetic endowment in an historical situation, and on the meanings a human being has assigned to them in the course of seeking a successful adaptation (or, at the least, in trying to secure against failure to adapt). Adler (1964) identified the field of study of Individual Psychology as located in "the relationship of the individual to the problems of the outside world." He went on to say:

> The Individual Psychologist must observe how a particular individual relates himself to the outside world. This outside world includes the individual's own body, his bodily functions, and the functions of his mind. He does not relate himself to the outside world in a predetermined manner as is often assumed. He relates himself always according to his own interpretation of himself and of his present problem. His limits are not only the common human limits, but also the limits which he has set himself. It is neither heredity nor environment which determines his relationship to the outside. Heredity only endows him with certain abilities. Environment only gives him certain impressions. These abilities and impressions, and the manner in which he "experiences" them — that is to say, the interpretation he makes of these experiences — are the bricks which he uses in his own "creative" way in building up his attitude toward life. It is his individual way of using these bricks, or in other words his attitude toward life, which determines this relationship to the outside world. (pp. 205-206)

We must attend both to the physical material of human life and to the social world apart from which it cannot flourish, and could not arise. To avoid errors we do well to consider Adler's image of the bricks: Bricks do not determine the kind of structures we build out of them. Their availability cannot account for the emergence of Georgian architecture in the eighteenth century. Artists and artisans decide these questions. Individual Psychology inquires into genetic endowment and environmental opportunity, therefore, as archeology may inquire into the quality of baked clay and the skills required in bricklaying. The purpose there is to guess at how architects in a given historical situation might have experienced the limits of structural possibilities, and how they achieved whatever seemed possible to them in the light of their understanding. Our purpose here is to guess at how children, in their creativity, have assessed the limits of biological possibilities, and how, by their lights, they accomplish what they believe is open to them within the compass of their assessments.

Genetic Possibility

According to Adler (1964):

> The style of life, in our experience, is developed in earliest childhood. In this the innate state of the body has the greatest influence. In his initial movements and functions, the child experiences the validity of his bodily organs. He experiences it but is far from having words or concepts for it. Since the way in which the environment approaches the child differs in each case, what the child feels about his capacities remains permanently unknown. With great caution, on the basis of statistical probability, we may conclude that where an organ inferiority is present, the child has the experience of being overburdened at the beginning of his life. How the child attempts to cope with this overburdening can only be discovered from his movements and trials. Here every causal consideration is in vain. Here the creative power of the child becomes effective. He strives within the incalculable realm of his possibilities. From trial and error a training results for the child and a general way towards a goal of perfection which appears to offer him fulfillment. (pp. 186-187)

Most accounts report that Adler's interest in psychiatry began to develop almost as soon as he had established himself in the practice of medicine. His office was close to Vienna's *Prater* (a kind of amusement park, zoo, and circus combination), and he had acrobats and other performers among his patients. People who earned their livings as exemplars of physical fitness confided their medical histories and weaknesses to him. Adler used the familiar medical concept of compensation as an organizing idea for understanding his new clientele. In medical usage, compensation refers to the overdevelopment of one organ or organ part to make up for damage or loss of function in another. He found it reasonable to extend the idea from an organ to the whole organism; thus it was possible to consider how an acrobat could be compensating for specific constitutional limitations by overall development and training of agility.

Adler's personal history provided a context for his ideas about felt weakness and the compensatory quest for strength. He could remember lying ill in childhood and hearing a physician tell his parents that there was nothing to be done to save his life. He remembered that his father had refused to give up, dismissed the pessimist, and called in another physician, who was willing to continue treatment. He also remembered the unwelcome sense of constriction he took from his bandaged legs as he endured a treatment for rickets. The meanings he gave to these experiences were expressed in his later observation on overcoming: "To be a human being means to possess a feeling of inferiority which constantly presses toward its own conquest" (p. 116).

We can see how Individual Psychology originated in the personal and professional interests Adler took in the various psychical compensations individuals erect against organic inferiorities. In considering general efforts to resist limits and overcome shortcomings, he came to locate the only true compensation for personal weakness and insignificance in a fully developed sense of our participation with others in shaping the common life (G., *Gemeinschaftsgefühl*). Individual Psychology did not abandon its original ideas in the course of its development. It still attends to any sense of smallness, incompleteness, or disadvantage in a person's self-awareness; it still notes the value of those images of strength, triumph, and perfection with which we console ourselves, and by which we inspire ourselves to persevere.

Our earliest awareness of ourselves arises while we still have much of the long period of infancy and childhood ahead of us, and are barely at the beginning of our physical and mental development. To know oneself is, at first, to know that one still has a long way to go. That we are responsive to stories of successes in overcoming difficulty may be owing to this common feature of our individual subjectivities. From ancient times, our pedagogy has treasured the story of Demosthenes, the stammerer honored for his oratory. In modern times, Scott Hamilton's story offered the same kind of encouragement. A congenital disorder of the pancreas, Schwachmann's Syndrome, marred his early childhood development, leaving the growth of his body irremediably stunted. It did not, however, crush his spirit as he perfected the skills that made him an Olympic champion in figure skating.

These stories mean, even to those whose development and constitution are unimpaired, that courage overcomes handicaps, that attitude can defy limitations. Courage is the readiness to act on the useful side of life, to contribute what we can to the solution of the problems we encounter in our situation as it is. In courage we press on with whatever we have at our disposal to face whatever challenges we meet. The courageous child says, "I am strong enough now for what is required of me, and I will be stronger still in time." The discouraged child changes this to, "I am never sure that I can be as strong as I must be." The conclusion for the former is, "Therefore, I will do what I can, toward the goal of success." For the latter it is, "I must do whatever I can to guard myself against defeat."

The Right Reverend Angus Dun died in retirement after a long life, his distinguished career having culminated in his position as Bishop of Washington, D.C. As a forerunner of the Civil Rights movement, he had used his position (and the pulpit of the National Cathedral) to challenge the consciences of political leaders on the demand for equal justice and for the removal of racist barriers to its attainment. (For these efforts his opponents had dubbed him "Black Angus," an attempt at a slur which he had accepted as an honor.)

After noting his many contributions, newspaper obituaries described some of the circumstances of his childhood: He had been born with a defect of the hip, which kept him from full participation in boyhood activities. The sting of this developmental disability was ameliorated by his knowledge that the condition could be surgically repaired after his growth had reached a certain point. At thirteen years of age he underwent the promised surgery, but while still in recovery he was stricken with poliomyelitis. After the fever had passed his mother came to his bedside and proposed an agreement to him. She asked that, whatever else might happen, they would agree never to talk about things he could not do, but only about the things he could do.

With this challenge she stimulated her son's courage. The story is an illustration of the principle of encouragement upon which so much in life (and in any therapeutic process) depends. If encouragement is lacking, a child's native sense of smallness can become an alien feeling of inadequacy. Dun's mother showed a confidence in him that called up his confidence in himself. Had she pitied him instead, he could have concluded, "Since I am so weak, I cannot count on myself. Therefore I must make sure that the people I do count on are always aware of how much I need them."

To inquire into the issue of genetic possibility is therefore to seek out the foundations of an individual's sense of "having what it takes." Certain probabilities do, of course, exist with respect to various conditions. Ugliness is not an advantage, for example. The loss of tonus in the facial muscles (which so often accompanies and expresses the experience of being rebuffed) further exacerbates its effects. The face of a bewildered and frightened child may give others the impression of someone wanting to withdraw. This may provoke a counter-withdrawal by them, leaving the child confirmed in the sense of being excluded.

Children who suffer faulty hearing (an impairment of which they can have no direct awareness) often come to think of themselves as slow-witted; others seem to them to be quicker to catch on to jokes and better able to follow along in exchanges of banter. It is a short step from this to suspicions that the others are deliberately keeping things from me, and not a much longer step to paranoid convictions that they don't want me to hear because they are talking about (and against) me.

Meanings assigned to body size and form however, or to any genetic patterns, are value judgments of such a personal kind that every probability can be defied by the single case, and the investigator can be startled by what is disclosed.

We will all sympathize with a complaint about a bodily injury or anomaly experienced in childhood as a deformity. It is easy for most of us to understand what is meant or implied by statements such as, "I was (too) fat," "I was (too) skinny," "I was (too) short," "I was (too) tall." It may be harder for some of us to see what is meant by, "I was too smart." ("It led the adults to expect more of me, and kept me from being accepted by the other kids.") It may be hardest of all for us to understand the complaint, "I was forever being told how beautiful I was." ("No one was ever interested in my mind, only my body because everyone *knows* that 'beautiful' means 'beautiful but dumb.'")

In any case, it is never wise to assume that you know how a child must have felt about any particular feature of the genetic endowment. Private sense is by definition elusive. It is important to be alert to things which we would not have expected, and to maintain respect for the unique ingenuity of this one person's continuing way of working out a successful adaptation. In Adler's (1964) words:

> It goes without saying that the inheritance, primarily of physical attributes but also of specifically human potentialities, is not subject to any doubt whatsoever. But what matters ultimately is what the child, the individual, does with the equipment he inherits. That is his own creative achievement. It is certainly human chromosomes, the individual's physical endowment, his endocrine glands, and

his blood constituents, which enable him to move, to function, and to look like a human being. But how and for what purpose or objective he molds and shapes these various formative factors becomes manifest only through the millions of variations in individual styles of life. (pp. 206-207)

Below are four brief statements by different clients, provided in answer to Item 31 in the *IPCW*, which inquires into childhood bodily development. These examples are followed by "The Case of Maria," whose childhood can serve as a paradigm for the positive uses of adversity.

Shortness of Stature. This man was the baby of the family, much younger than the youngest of his siblings. He stated:

I was short for my age. I had a sense of defenselessness as a child because of it. This improved somewhat when I reached puberty and kept up more with my peers. Taller guys, though, always seemed to be better accepted. It felt like a burden to me to have to *do* something to be accepted.

For this man, shortness of stature served as a rationalization for his feelings of vulnerability, to which he expected others to respond with caretaking. He had trained for this, first in his family of origin among the much older siblings, and later in life with friends and lovers. His failure to form a lasting attachment with a sexual partner was ascribed by him to his disadvantages. He judged life to be too hard, and believed that he shouldn't be expected to make his place by way of making contributions. To have to *do* something to be accepted was a burden, out of line with the expectation that he could gain a place by *being* something (short) or *feeling* something (defenseless).

Physical Strength, Coordination, Speed. The younger of two children, a woman with an accomplished older brother (the favorite of both parents) said:

> I always felt good about my body. I was agile, and learned how to do things quickly. My good coordination was always commented on by people. I liked tumbling, and I could throw a ball well. I was a very good runner. I learned pretty early that I could excel over my brother in these things. I was always challenging him and showing him up. He hated it that I ran faster than he did.

This woman's physical ability was acknowledged and encouraged early in life, and she developed herself along the line of this effectiveness. She realized that in this realm she had an advantage over her brother, and could win a place with her parents in spite of their favoring him. She competed successfully in several sports as she grew older. As an adult she continued to press her advantage, seeking out men she could defeat, both on the playing field and in professional life. She came to see us, wondering why she had failed to make a match with a man worthy of her respect.

Ethnic/Racial Identification.   A laterborn son in a large family reported:

> I didn't have any problems with my body. I was about average in my abilities and could hold my own. But, you could tell which side of my parents' mixed marriage I came from! My mother and father had no trouble with this, but my mom's family, being white, didn't have much to do with us. I didn't like my relatives on her side. They seemed to look down on my dad, and I didn't like that. I was the only one of the kids who was really black, and so it was easier for me to identify with my dad. That suited me. I think he's a great guy.

Here a positive evaluation of dark skin color was made by the boy in defiance of the norms of his mother's family and the larger white community. His solidarity with his admired father compensated for any disapproval or difficulty he might encounter as a consequence of being black. He regarded what someone else might have considered to be a detriment to be a mark of distinction in his family and elsewhere. He and his wife came to see us because she complained of his lack of sympathy. She was struggling to make a place for herself in a male-dominated profession, against her uneasiness about the incompetence of women as defined by her feminine guiding line. Her self-doubts were unintelligible to him. He had not allowed prejudices to stop him, and he responded to her concerns with pep talks that sounded to her like criticism, and left her feeling worse.

Weight. A woman with two younger sisters said:

> I was born fat. No, really, I was always a chubby little kid, not fat, but somewhat overweight. I hated it. My mom cared a lot about the way we looked, and she was always on my case about my weight, always trying to get me on a diet, commenting on everything I put in my mouth. Mealtimes were awful. And my sisters got into it too. They'd tell her if they saw me eating stuff I wasn't supposed to have, like candy bars, that sort of thing.

Food issues are often the *stated* topic of power struggles between mothers and daughters, though the *subtext* is typically a fight over who is in control of the daughter's life (i.e., body). When such a fight is not resolved quickly, the daughter can develop a full-blown practice of obesity or anorexia nervosa, with or without bulimia. In the case cited, the client's genetic disposition enabled her to put on weight and to maintain it. She made use of this capacity to fight her mother's control, as if to say, "You may dictate to me about everything else in my life, but you cannot control what I put into my mouth, what goes down, or whether it stays down." She said that she hated the extra weight and, of course, she did. It was the high price she believed she had to pay in personal unhappiness in order to defeat her mother. She was in her forties when she decided to make choices for her life unrelated to what her mother might or might not think about them.

The Case of Maria

Maria was the second of five children of the second marriage for each of her parents. The first of these children was a boy, as was the third, and her father had two much older sons from his previous marriage. She was therefore her father's first, and for some time, only daughter. Mother had one child from her former marriage, a pampered daughter, eight years older than Maria.

Maria was asthmatic, knew that her parents did not expect her to live, and did not expect herself to live. She perceived herself as frail and sickly all through childhood. Now she is the mother of four grown children, and still regards herself as vulnerable to illness unless she takes every sensible health precaution available to her (good diet, exercise, adequate sleep, etc.).

Maria had ample opportunity to make a place for herself as the pampered pet of her father, and to set a course for a life as a neurotic invalid, demanding special attention and service from others.

However, her hidden reasoning about herself as a child was that, since she was soon to die, every day was a bonus in which to discover how much she could do. Recognizing the special place she had with her father, she did her best to demonstrate to him, in return, that she was a responsible and reliable daughter, even if she was doomed to an unavoidably early death. Her older half-sister's self-indulgent style did not meet Father's approval, and, in cementing her place with him, Maria felt all the better able to contrast her more sensible and practical approach to her sister's frivolities.

The older sister also assisted her by refusing to cater to her crying, and by persisting in teasing her, possibly out of resentment at being displaced as a babied pet. Maria got angry with her in response to the unrelenting teasing, and unexpectedly won her sister's respect by standing up to her.

Maria's own children are all self-reliant and strongly independent. Believing that she would not live to see them grow up, she stressed the importance of their being able to care for themselves.

Maria has a senior management position in a large corporation, and entrusts a great deal of initiative to her subordinates, who "always know exactly where things stand, and what has to be done," whenever she is required to be away from her office.

So far, we have the story of Maria's courageous and successful rebuff to the threat to her life she experienced in childhood. What, then, brought her into psychotherapy? She had been married twice. Both alliances ended in divorce. She said of her husbands that they offended her by acting as if they should be able to control her, which she found intolerable. Her father had never done anything like that; he was happy just to have her alive.

Environmental Opportunities

Individual Psychology has been misunderstood to be an environmentalistic theory, tracing the entire psychological structure of a person to environmental influences. To examine this misunderstanding more closely, one would have to ask: Who seeks, who answers, who utilizes the impressions from the environment? Is man a dictaphone or a machine? There must be something else at play. (Adler, p. 207)

In the course of family counseling, parents often complain that a child takes advantage of the parents' generosity, good nature, preoccupation with maintaining a tidy house, or whatever. In response, we say that bright people always take advantage of their advantages.

Lifestyle assessment can be thought of as an ethological study of the human being in relation to the human habitat. Any consideration we give to the importance of the environment in which an individual grew up, therefore, must avoid crude assumptions of cause and effect. At the same time, we must maintain an awareness of probabilities. There is a real difference between growing up in the only Roman Catholic family in a rural county otherwise populated entirely by Protestants, and being part of an urban family in a neighborhood in which everyone was Catholic and belonged to the same parish church.

There is also a likelihood that familiarity with the court system will be as common among the poor as academic degrees are among the affluent. But, to assume the inevitability of dysfunction among the poor is (for the poor) to indulge in self-pity, and (for everyone else) to condescend; to be surprised by the occurrence of dysfunction among the privileged is (for the privileged) to indulge in self-righteousness, and (for everyone else) to fawn. *No probability can be relied upon to be true in the single case.*

Remember figure and ground, and the context in which uniqueness is recognized. The understanding of the individual variant is our goal. We must not lose sight of the creative power of any child to defy the statistical odds, or to make a fresh and positive meaning of a situation that is unpromising. Nor should we overlook the power of a child to make of a good situation with every objective advantage a lifeless and negative meaning that becomes the ground for resentment and complaint.

The Case of Lloyd

Lloyd grew up in a poor ethnic neighborhood that had lost its coherence to urban development and economic hardship. Mother had a part-time job in a manufacturing plant. Father worked only occasionally, and when he did work was regularly fired for drinking. Lloyd remembers having to take care of his father when he had been drinking, and being embarrassed by having to go get Father and bring him home from a nearby bar.

Lloyd spent as much time as possible with his maternal grandmother, a widow whose husband had died in an industrial accident years before. She made do on a tiny pension. His other grandmother was dead, and the paternal grandfather had abandoned the family when Lloyd's father was a child. Other family members had scattered. He describes the parental relationship as violent. He has no memory of any affection between the parents. "He'd drink, and she'd bitch at him, and he'd hit her. I was terrified of him, and I felt sorry for her."

He was aware that a few blocks away there was a nice neighborhood where the houses were well-maintained and the children well-dressed. "They wouldn't have anything to do with us. Most of them went to the parochial school, and we couldn't afford to go." Lloyd went to church in the parish with his mother, but attended the overcrowded, rundown public school.

He says that he made up his mind, "almost as soon as I can remember, that I was going to get out of there if it was the last thing I ever did. It was so depressing. Even my grandmother's situation, which was better than most! There she was, this little old lady, and she didn't have anything. I knew I'd make money some day and take care of her and my mom — and I did."

Lloyd went into banking and made his way from clerk to president; he had known from childhood he wanted to make money, and he did. Adler observed, "[The] artful construction of [the individual's] form of life . . . follows by no means a causal process. The decisive factor is always the concretized fictional goal of life" (p. 282).

Lloyd came to see us because he was depressed. "I don't find life interesting any more. I work hard, but what does it get me? I've got lots of money, but I don't enjoy it any more."

It turned out that Lloyd had made taking care of women his life's work. He had a beautiful house in the suburbs, complete with a gorgeous wife and two well-groomed daughters in private schools.

He pampered all of them, and since they felt entitled to what he provided, they responded with demands and petulant insistence on having their way, with the result that he felt unappreciated.

He had overcome a background of poverty, alcoholism, and brutality. He had also developed some mistaken ideas, especially about money and relationships of intimacy, which he reconsidered in psychotherapy.

The Case of Christine

Christine, a thirty-two-year-old Caucasian woman with two children entered therapy for the first time. Her nine-year marriage to Stephen had come to an end the month before, when he announced that he had fallen in love with an associate in his law practice and wanted to marry her. He had moved out, and Christine knew his mind was closed to the possibility of reconciliation.

Christine came from a prominent family, known for its business interests, money, accomplishment, and good works over three generations. When her mother had become pregnant with Christine, it was a shock to the parents, who thought their childbearing years were over. They had one son who was fifteen when Christine was born. Like most parents of their generation, they decided to make the best of it, and were delighted with their little daughter.

The family lived in a wealthy suburb of the large city where Christine's father had his office. He traveled often, and had many civic responsibilities. Mother was equally busy in her own domain. She served on the boards of several charities and was the hostess at frequent social events. Both parents played tennis and golf, enjoyed their friends, and, when possible, took trips together.

Christine described her family life as taking place "in a goldfish bowl." The pressure to be proper and to do the right thing was experienced as "immense" by her.

When the family was together she saw her father as "gregarious and affectionate" and her mother as "motherly, almost grandmotherly." She said that her brother adored her: "He treated me like a little doll. When my parents entertained, he'd trot me out and show me off. They'd say, 'Isn't she cute!' and he'd whisk me away again. It was fun."

Christine and her brother were cared for by a staff of indulgent servants. She said, "I really had pretty much whatever I wanted, as long as I was good. They picked up after me, took me places, and bought me things."

Christine went to boarding schools and camps, and often felt misfitted. She complained, "I never felt comfortable with children my age. I don't know what was the matter with them. We never seemed to have anything in common. I ended up pretty much staying to myself in school." She found her refuge in academics. She was brainy, got top grades, and wanted to be a scholar. She went to Vassar College, where she graduated with highest honors, then to Harvard to get her Ph.D.

It was at Harvard that she met Stephen, who was in law school. "He was my first love. We had a mutual interest in the Elizabethan period. I was studying Shakespeare, and he was studying the Common Law. We got married in graduate school. I think that my parents were relieved that I had found someone."

She said that their marital problems began soon after they completed school. "We moved here, bought a house, and he began his practice. It wasn't long before he began spending a lot of time downtown. I knew he was establishing himself, but I resented his not spending more time with me. I just felt that it was more important to be together, since money was not a problem, but he was involved in the practice, and quickly got into a lot of civic things, like my father.

"Then, before any of that got resolved, we had the two children, one right after the other, and I got frantic. I couldn't do anything.

Stephen got really mad at me. I remember one time he said, 'We aren't playing house, for God's sake!' I was hurt by that, really hurt. I knew we weren't playing house! He didn't see how hard it was for me."

Christine thought the solution was to hire help, but after going through a number of bad experiences, with much complaining, Stephen finally put his foot down. Christine reported, "He didn't want them around, intruding on his privacy and causing problems. He said he'd rather do it himself. And he did. He shopped, cooked, everything! I ended up staying out of the way."

There was a running battle between them during this period over his desire to have the family move downtown, closer to his work. "I said, 'Absolutely not!' Can you imagine? What kind of life would that be for children?"

Christine had known many advantages in her growing up years. They included money, position, good health care, an excellent education, cultural opportunities, a family that doted on her, and the example of socially concerned people. With all of this, she had failed in the tasks that confront everyone: friendship, work, and love.

The first clue to her lack of preparation for life was shown in her report of her inability to make friends among her peers; second, was her failure to find something useful to engage in while Stephen was establishing himself in his life work; third, she failed when she found herself in the position of being responsible for others, namely her children. When things did not go her way, she retreated, with complaints and resentment. The loss of her marriage was one of the outcomes.

What had gone wrong? She had made a mistaken conclusion about herself and about life. She believed that she *required* the services, support, and attention of others in order to live, and that aside from these she did not know how to proceed. Her belief was that it was up to the others to take care of things generally, and to take care of her in particular. When they failed to do so she felt, for all the wealth of her background, impoverished.

Conclusion

In this chapter we emphasized the significance of the child's own evaluation of childhood circumstance relating to questions of heredity and environment. We have suggested that the art of guessing depends on a use of probability as a framework in which one is often surprised. We have cautioned against the uselessness (and danger) of causal and stereotyped thinking. Nothing believed to influence the individual can be assumed to be true unless the individual asserts it. In matters of lifestyle interpretation, *the client is sovereign.*

In Chapter 4, among topics suggested for examination in a lifestyle assessment, we listed "other particularities," including (but not limited to) the ethnic/racial, religious, social, and economic situation of the child. In the next chapter we continue our review of family material in considering these particularities in the framework of the child's training and self-training for life outside the family (especially in the neighborhood and school) and for meeting the challenge of adolescence.

Exercises for Chapter 9.
Genetic Possibility and Environmental Opportunity
(Use in dyads, small groups, or class, or with clients.)

Genetic Possibility

In a few paragraphs, describe your sense of your physical self in the childhood years up to age nine or ten, as to your advantages and disadvantages regarding your physical constitution, including mental acuity. Then, consider and describe your presentation of yourself to others. How did you think others perceived you at that time?

Environmental Opportunity

In a few paragraphs, given the environmental circumstances in which you developed as a child, describe what you thought at that time was *open to you to do in life*, then describe any *limits* you believed existed or believed you would encounter as you moved forward in your situation toward adolescence and adulthood.

Chapter 10. PARTICULARITIES OF THE
CHILDHOOD EXPERIENCE AND THE
CHALLENGE OF ADOLESCENCE

Throughout the lifestyle inquiry, there are many opportunities to note and record information on particular ethnic/racial, religious, socioeconomic, and other matters characterizing a client's childhood experience, especially while taking the descriptions of parents and inquiring into the histories of grandparents and other members of the extended family. Also, if there are other details of importance to the client, the *IPCW* provides a further opportunity to identify these by specifically asking for such information in Item 13, Family Milieu.

In Chapter 4, we recounted the entire inquiry in "The Case of Dan." In this chapter we will return to that material in order to draw attention to the particularities of Dan's childhood situation discovered there. We will also review "The Case of Eugene" as an example of the way this information permeates the discussion of the family, rather than standing apart as discrete subject matter.

Notice that we are not so much interested in the *facts* of Dan's or Eugene's childhood milieu as we are in the *meanings* they each gave to the facts and the *uses* they made of them. We cannot repeat often enough that, "We are influenced not by facts but by our opinion of facts" (Adler, p. 192).

To emphasize his teleological analysis of psychological phenomena, Adler (pp. 205-206) taught a "psychology of use" (G., *Gebrauchpsychologie*), distinguishing it from a "psychology of possession" (G., *Besitzpsychologie*). Individual Psychology does not address what one has (traits) so much as what one does with what one has. We study the particularities of a childhood situation to learn how the person either turned them to advantage, or shifted responsibility to them as rationale or alibi for failures or setbacks.

In the cases presented here we consider these major topics: ethnic/racial origin, religious practice, social situation, and economic status as these are interpreted by the client.

Adler's emphasis on "social embeddedness" (p. 126) prepares us to consider how ethnic/racial evaluations vary according to age-groups. In the United States older persons can remember (and may still retain) feelings of embarrassment regarding mixed-race ancestry. It will not be unusual for an older person to report a general unwillingness to discuss or even admit to such ancestry, however remote a racial mixture may be.

> - A white woman of seventy-six told us in 2010 that she was a direct descendant of John Rolfe and the Native-American princess, Pocahontas, who lived *four centuries ago.* Her mother passed this information on to her as a family secret, not to be mentioned in the presence of her grandmother, for whom it was a family disgrace.

It is unlikely that such attitudes will be found among young adults today. An extensive front page report in *The New York Times* (Saulny, 2011) profiled the activities of the Multiracial and Biracial Student Association at the University of Maryland as exemplary of a rapidly-changing demographic. Of special interest for therapists alert to the individual variant in any population is the variety of meanings these students gave to their personal experiences of mixed ancestry, all tending toward the positive, good-humored, and mutually respectful.

Since innovation in social practices and attitudes tend to travel upward from younger to older age cohorts, we can expect to encounter attitudes in younger persons that may even challenge the counselor's unexamined biases.

Sometimes we can be surprised by joking self-parodies of ethnic stereotypes:

- Michelle, a successful small-business owner, a woman in middle life, said, "I'm half Irish and half German. I like to drink and clean my house."

The counselor must also be alert to divisions regarding ethnicity that are often painful, reflecting the individual ways children have of perceiving and evaluating any particularity.

- Inez, a firstborn daughter, strongly identified with her ethnic heritage, engrossing herself in a study of the family's history, becoming proficient in the ancestral language, and engaging in the activities of the family's ethnic fraternal organization and religious practice. Her younger sister, on the other hand, showed no interest in the family's ethnic background and activities, learned only a few phrases of the "kitchen" version of the parents' original language, and further separated herself from the others by exaggerating her way of being a "typical" American.

Particularities in
The Case of Dan

Ethnic/Racial Origin. Dan reported that his parents were both of English ancestry, adding that his mother "said her people were of 'better stock' than the Allens because they'd been here since colonial times, and the Allens hadn't come over until 1885." Later he said, "I can't be all that sure about Dad's family; he *says* they're English, but Mother's *is* English, through and through.

Religious Practice. Dan said, "I was steeped in the Episcopal Church. Sang in the choir and so forth, although it dawned on me as I got older that we were a little out of our league there, in terms of class. But it was part of Mother's thing. She always referred to us as 'Anglicans'! My father didn't go, by the way — another bone of contention between them, and another opportunity for her to lord it over him."

Social Situation. "Mother was very ambitious for me. . . . She was determined that I wasn't going to be deprived of the opportunities that came with higher education." Mother also wanted a higher station in life for her sons: "She demanded that I look spotless, well-groomed, and that I was perfectly behaved." Her response to the accident in which Dan broke two of his permanent teeth was to be "terribly upset" in her concern for his appearance, not his well-being: "You'll be disfigured for life!" Finally, we know that she didn't want Dan to play with the neighborhood children, whom she considered too rough.

Economic Status. While they were "better off" than the others in their neighborhood, this was a poor, working-class area. Mother complained that there was not enough money, blaming this on Father's gambling. As compared with others, though, "our house and yard looked nicer, we ate better, and I was better dressed," which tells us either that Mother was a good manager of what was available, or there was more available than she recognized as ample, because of the extent of her ambitions.

Discussion

We concluded that in Dan's family, where Mother was the dominant figure, Dan felt obliged to follow her lead in claiming the socially preferred background, and to honor her social ambitions and goals. Mother aspired to see her boys "rise above" their beginnings: They should get a good education, learn and maintain propriety (good manners, good clothes, good teeth, good Anglican training), in order to better themselves and escape into the middle class, as she had been unable to do.

How did these values and attitudes serve Dan when he went out into the neighborhood and school to make a place for himself among others? We can see that in the neighborhood Dan took the position of being better than the other children, and that this carried over into the school, where he studied and was bright enough to be taken out of the neighborhood group and elevated to a higher track, again showing himself to be better than others. There was, however, an ironic price to be paid: Having located himself in this vertical frame of reference, he found himself out of his element and felt "ill at ease." His speech gave him away as unacceptable, and he was publicly identified as not as good as the others.

About his dealings with the other sex, Dan said, "When I hit puberty I ran into trouble with girls." He didn't date until he was eighteen, and his first intercourse was when he was twenty-five. His caution in this related first to his apprehension regarding the power of women (his feminine guiding line), and second to his education in propriety (sex was not "right") and appearances ("I was unattractive"). Finally, he was also "terrified" of getting someone pregnant and thereby derailing himself from the completion of his education and the route up and out, in keeping with the ambitions he had accepted as his own. With his view of women as knowing better, Dan was vulnerable to being caught up in the life of one of them, which is not the same thing as being a man who makes a life with a woman.

About working, Dan said that he "didn't have a clue" about what to do when he grew up (there were not many options available to his imagination from the world of work engaging the men in his family and neighborhood). He was drawn to the idea of being a railroad engineer, but knew that "it wouldn't have been the right job . . . [since] it was manual labor." The only thing that mattered was, "just do the job," and any job would do, except that it had to be in an office. In avoidance of a decision Dan stayed in graduate school for many years and was finally drawn into the world of the laboratory, where he was able to satisfy the demands of his ambition without actually being in an office.

We can see that Dan, having learned Mother's lesson, felt Mother's distress as a consequence. If the world was divided into two groups, the Better and the Worse, and if being best in the Worse group was not good enough, he would strive for a place in the Better group — and always wonder whether he was the worst in that one, even at his best.

The Case of Eugene

Eugene grew up as the much youngest of six children in an immigrant Arab-American family. It was an Eastern Christian family that had lost its ties to the Arab-Christian community, with no new religious ties to put in their place in a predominantly Roman Catholic working-class neighborhood in a small town in Ohio. It was a family determined to escape its immigrant status by working for financial success, and the independence that would bring, while at the same time keeping very much to itself. (Eugene's name was chosen by one of his older brothers, who convinced his parents that it was a typical American name.)

Father was described as a bully at home and a person who was accommodating to others to the point of servility outside the home. Mother was described as caring, friendly, and cautious. The parents spoke little English. Father had worked his way up through three jobs during the time of Eugene's growing up, and Mother worked at various part-time jobs whenever it was possible for her to leave the children, who were her first responsibility. Mother's business sense enabled the family, in time, to buy a small grocery store and establish itself. Eugene saw Mother as "the brains" of the family and Father as "the brawn." Being the last of the children, Eugene benefited most from the family's growing financial success and considered himself to be spoiled "materially, but not in terms of permissiveness. Father saw to that."

Socially, the family had a place among the few Arab relatives who lived nearby, but had a "low social standing" otherwise. Eugene said, "I was aware that my family was somehow different, and that we weren't seen as 'real' Americans. This was a problem for me. I was ashamed that my parents spoke so little English, and that we weren't invited into other people's homes. It's interesting that *they* came to our house when we lived in back of the store (that is, my father's cronies did, especially on Sundays when the taverns were closed and they wanted to mooch a drink), but we never went to theirs. This bothered me. I definitely had the feeling that we were below other people socially, but I also had the feeling that this would change when we had money, because that was what people were interested in." Ethnicity, religion, and social standing all paled in comparison to the goal of economic security. All their other difficulties would be overcome if only the family gained financial independence through their hard work, and kept a "clean record," which meant avoiding anything that might come to the attention of anyone in authority: "Our motto should have been, 'Caution: Don't Make Waves!'"

In the neighborhood Eugene was a follower. "I went along with what the other kids wanted to do. I was fast and strong, so I was picked for teams, though I was never a star or anything like that. I was an average kid; I had a reputation for being a really nice kid." So, even though he felt defined as different from the others, he made his place among them by his strategy, based on the family values, of going along and working hard (in this case, at playing ball, in both senses of the phrase).

At school he was bright enough. His only handicap was his belief that his parents represented a handicap. "I didn't want them involved in anything at school, because of their poor English, so I was very careful not to do anything that would result in their being called in. In this way he imitated his parents, avoiding the attention of the authorities by keeping a clean record of his own in his own world of the school. (Eugene was the first of the four children to attend college, and the first to graduate. He went on to become a businessman — like his mother, but in real estate.)

Consider now how he made use of his training and self-training for meeting the challenges of adolescence: adjusting to puberty, learning how to be intimate with someone of the other sex, and meeting the economic demand of work.

"I was very bashful with girls. I felt I just wasn't good-looking enough to ask them out. Whatever I learned about sex, I learned on the playground from the other boys. Also, one of my older brothers tried to fill me in, but that was pretty useless, and in fact a little scary. Actually, in those years, the girls and the boys were quite separate. Maybe it was the era, maybe it was the town where I grew up, but we just weren't together very much. All through my junior and senior high school years, I never went out on a date. That seems incredible to me now!" He had his first sexual experience in high school when he went along with a friend to pick up a girl who "had a bad reputation"; he and his friend each had intercourse with her. He said, "It scared the hell out of me. I didn't know what would happen. I wasn't too pleased with what I had done." He said that later on, "My wife courted me, or I might never have gotten married!" Considering Eugene's place as the much youngest child in his family, it is not surprising that he did not take the initiative in establishing an intimate bond with a woman, instead allowing himself to be chosen by her. As he put it, "I always went along with what the others wanted to do."

About his choice of career, Eugene reported, "I always wanted to be rich, but I didn't know how to do it. My career just kind of grew. My mother wanted me to be a doctor, but I didn't want to go to school that long. I don't know. Looking back, it seems that whatever happened to me just happened, without any particular plan on my part."

In the same way as he experienced himself as being chosen by a woman, Eugene saw himself as chosen by various employers, including a person with whom his mother had had some real-estate dealings, so that little by little he had found himself occupied in that field. His uncontentious and ingratiating way of being with others stood him in good stead, and may have added to his success.

Getting along, not making waves, and keeping his eye on whatever opportunities chance put in his way were practiced operations for Eugene, and they worked for him in gaining a wife and gaining wealth. They failed him whenever he was called upon to oppose anyone, be decisive (especially with respect to an unpopular stand), manage and direct the activities of others, or make and carry out plans. Further, the idea of being "different" was anathema to him and so placed a firm limitation on his creativity.

A theme of Eugene's life as the last and (during childhood) the smallest in his family was, "I don't want to be left behind," and his compensation against this danger was "to get ahead, or — at the very least — to stay even." The crisis that brought him to consult with us came when his partnership broke up in an acrimonious public dispute, and many of his "best people" left him to go into another firm with his former partner, where they believed they would enjoy (as one of them put it) "an innovative edge."

Particularities in
The Case of Eugene

Ethnic/Racial Origin. Eugene's story features his ethnic difference from others in his community. In working with us he often made reference to his having been a member of the only Arab family in town. He also emphasized that he was part of an immigrant family, and was aware of his first-generation status, of which his "typical American name" was evidence. By contrast, he allowed himself to be embarrassed by his parents' imperfect mastery of English.

Religious Practice. The family left its religious life behind in coming to the United States. Their Eastern Orthodox customs differed from the Roman Catholicism that dominated the working-class town in which he grew up. Once or twice a year they would make a pilgrimage to Cleveland for a major liturgical celebration, but this only left Eugene feeling further set apart from others in his own town.

Social Situation. The family had "a low social standing" compared to others. Eugene resented the fact that Father's cronies came to their house, but that members of his family were not invited to the houses of other people. This, combined with the "servility" he believed Father to exhibit generally, left Eugene with the idea that the family was not only looked down upon, but was exploited.

Economic Status. The parents' example of hard work and its rewards was in front of Eugene at all times, as was the belief that with money comes everything else. He was confident that money "was what people were interested in," and that the family's problems would be solved when it was established financially.

Discussion

Eugene overcame his feelings of being ethnically and religiously different from the others and being "less than" they were by accepting the standards of the community as he understood them, and by taking part. He had learned to cooperate inside the family, where everyone worked hard to achieve financial success and to maintain the family's anonymity in the eyes of any authority, group, or person who might stand in the way of that success. He carried this attitude into the world of the neighborhood and school, polishing his skills at "going along" to the extent of making a good marital alliance and a successful business career. Eugene's objective life situation had in it the potential for his coming to feel like an outcast, and for developing his resentments into an antisocial attitude; his subjective sense was, however, that he would be able to move beyond the limits of his immigrant family if only he made enough money, which he set out to do, and did. But, this did not move him beyond the limits of the cautious lifestyle he had practiced as a member of his family.

Conclusion

We conclude this chapter on "Other Particularities" by reiterating Adler's statement:

> Meanings are not determined by situations, but we determine ourselves by the meanings we give to situations.
> (p. 208)

Next, in Chapter 11, we present the process of review and interpretations of the early recollections, addressing special problems that arise in the course of this task, and providing a complete transcript of the process in "The Case of Janice."

Exercises for Chapter 10.

Particularities of the Childhood Experience
and the Challenge of Adolescence
(Use in dyads, small groups, or class, or with clients.)

Particularities of the Childhood Experience

In a brief sentence or two, state what each of the following descriptors meant to you as it could be applied to your family when you were a child, up to age nine or ten: (1) economic situation; (2) ethnic or racial origin; (3) religious attitudes and practice; (4) social status.

Concretization of the Goal

"The goal must be made concrete to become clearer" (p. 99).

- Think back to when you were little, ages three to six. What did you want to be when you grew up? What was there about this that appealed to you?

- When did this idea change (if it did)? If it changed, what new idea took its place? What was there about the new idea that appealed to you?

- Think about what you are doing now in life. See if you can discover how your present occupation relates to your earlier conceptions.

The Challenge of Adolescence

Write a few paragraphs on how the meanings you have ascribed above to the categories of economics, ethnicity/race, religious experience, and your place in society entered into your opinion of yourself in adolescence and how your evaluation of these circumstances affected your readiness to meet the new task of love and sexuality.

Chapter 11. EARLY RECOLLECTIONS

Memory, like attention, is selective, in keeping with the economy of the mind and the purposes of the individual. Actively to remember everything, even if it were possible, would so clutter and fill attention that it is difficult to imagine how we could endure it. The theoretical problem, therefore, is to account not for the fact that we forget so much (which is the burden of theories of *repression*), but rather for the fact that we remember what we do.

There were numberless events in the course of our childhood years, before we reached the age of nine or ten to which we attended and in which we took some interest at the time. In most cases, we are likely to retain images of only a few incidents, perhaps six, perhaps as many as ten, which we can call up and preview without conscious effort or any sense of purpose.

Individual Psychology is a theory of *expression*. It assumes that we retain these particular memories in order to maintain an orientation through time, to rehearse our understanding of the fundamental issues of life, and to provide ourselves with reminders of the reliability of our convictions. Cultures maintain continuity by means of myth and liturgy: "As it was in the beginning, is now, and ever shall be, world without end." Similarly, according to Powers (1973), individuals maintain identity by means of early recollections.

To understand this process requires that we recognize it as dynamic, seeing it as an active recollecting in the present moment of recall. Of all our movements, the activity of remembering is the most characteristic. It is in this activity that we reveal the style of our movement in purest form. If hyphenation could make our meaning clearer we would say that in our early re-collections we are re-constituting our world from its beginnings; re-membering the elements of our lives into the one body of our identity; re-calling assertions of meaning out of ourselves against the threats of confusion or despair; and re-viewing ourselves in our situations in our practiced way of looking at things.

To know your early recollections, therefore, is more than to have a familiarity with a collection of interesting anecdotes about your childhood. It is to have access to your basic schema of apperception, and to be able to see how you see yourself, others, and the world. It is to hold the key to the biased apparatus by means of which you are predisposed to perceive only the things that suit your purposes among all of life's possibilities, and to grasp only what confirms your convictions in any of life's information. It is to know how you have "made up your mind," and how you "keep in mind," and are "mindful of," those things that guide your choices and reduce your uncertainties. Finally, it is to know what you expect of yourself and others, and so to be able to predict how you will act in response to those expectations.

In your early recollections we will find your evaluations of yourself, your world, and the course of action open to you, as if in answer to the questions: What kind of a world is this? What kind of a person am I? What must a person such as I do in a world such as this to find a place of security (to fit in) and significance (to stand out)? What are my chances for success in all of this?

These are, if not the eternal questions, then the lifelong ones. We began shaping answers to them in the trials and errors of our earliest efforts. To the extent that these most primitive formulations worked to guide us, they became the bases for our further explorations, and for every further articulation of our personal values. If our lives are no longer satisfactory we will begin to reconsider the answers we have worked out. It will be hard. Our only way of thinking about them is itself a part of the way we have of doing everything, part of the way we have survived until now. Harder still, we cannot speak clearly about our basic convictions because they are not available to us except by implication in the images and stories of our recollections, which feel innocent of purpose or meaning. This is particularly true of our more extreme (and hence, more problematical) biases. Since they would not stand up to the scrutiny of commonsense, we must, to maintain them at all, keep them unarticulated, out of the public realm of language. We pay a price for keeping them un-understandable, of course: They must remain un-understood by us.

Basic convictions, especially to the extent that they are biased and private, are best maintained and expressed through metaphor, and best illustrated by the seemingly unassailable certainties of private experience. After all, most of us feel safe in believing that almost anything can happen once. Some of us are even willing to believe that our lives are special, and actively to remember unique things that happened to us, perhaps things that the rest of us would believe could not have happened, not even once. One client told us of his boyhood experience of levitation, as he walked alone through an open field. No one could have guessed by looking at him (and no one with whom he conducted his prosperous business did guess) that he was raging against paranoid delusions of being humiliated ("put down"), or that he was dangerously close to doing violent harm.

The skillful gathering, interpretation, and explication of a client's early recollections is a valued art in Individual Psychology. It is regarded by us as pivotal to the psychoclarity process of understanding lifestyle, and as close to essential as anything can be to any form of effective psychological intervention.

Adler (1964) put it this way:

Among all psychological expressions, some of the most revealing are the individual's memories. His memories are the reminders he carries about with him of his own limits and of the meaning of circumstances. There are no "chance memories": Out of the incalculable number of impressions which meet an individual, he chooses to remember only those which he feels, however darkly, to have a bearing on his situation. Thus his memories represent his "Story of My Life"; a story he repeats to himself to warn him or comfort him, to keep him concentrated on his goal, and to prepare him by means of past experiences, so that he will meet the future with an already tested style of action. (p. 351)

Early recollections take their place in the lifestyle assessment in three phases: Gathering and recording (see Chapter 4 for a discussion of this); review and interpretations (the subject of this chapter); and the early recollections summaries (Chapter 14).

In the remainder of this chapter we explain the procedure we use for the review and interpretations of early recollections. We then present a method for interpreting the recollections by attending to five key categories (context, content, gender, movement, and evaluation). Finally, we demonstrate the use of these categories in four examples. The chapter concludes with the complete transcript of the review and interpretations of the early recollections in "The Case of Janice."

Review and Interpretation of Early Recollections

After The *Summary of the Family Constellation* has been completed, the next task in the lifestyle assessment is the review and interpretation of the early recollections. This may follow immediately, or it may be scheduled separately for the next session.

The recording therapist reports each early recollection in turn, reading a verbatim record. The consulting therapist inquires into each recollection, and leads a discussion with the client, with the collaboration of the recording therapist. It is the consulting therapist's task to ascertain what the meaning of early recollection is to the client, by uncovering with the client the personal attitude, evaluation, or conviction the story expresses. To do this requires extending one's imagination in order to experience the memory as if it were one's own, until one sees the situation from the client's vantage. The object of the discussion is to discover how self, others, and the world looked to the client in the moment recalled.

When the consulting therapist succeeds in conveying an understanding of what the situation of the memory means, the client will exhibit some form of the recognition reflex (laughter, grinning, or blushing, with a pronounced meeting of the eyes) or will offer some verbal assent ("You've got it," or, "That's it!") after which the recording therapist reads the next recollection.

After reviewing each recollection in sequence, a second review follows in preparation for composing and recording the summaries. The recording therapist reminds the consulting therapist of the sequence of the recollections, of observations and clarifications noted in the first review, of the client's age at the time of each incident, and of any details that remain puzzling or problematical. At the end of this second review and discussion, the consulting therapist dictates two summaries of the interpretations based on them. The first is titled *The Pattern of Basic Convictions*; the second is titled *The Mistaken Ideas*. (See Chapter 14 for more information on the way in which these summaries are organized.)

Five Key Categories of Interpretation

For teaching purposes we have identified five categories as of special importance in interpreting Early recollections. Do not limit your vision. Empathy will go beyond categories to a sense of the whole world of another's experience; still, it should not fall short of or ignore the following:

1. *Context.* In the most general sense, the context of any recollection is the world itself, which presents itself as an ambiguous stimulus to which the recollection stands as a record of projective response (Mosak, 1958). (Poet Allen Ginsberg once said that we do not see things as they are, but as we are.) Any recollection could, therefore, be interpreted to some extent by itself, in an academic exercise apart from knowledge of the person recalling it. This would be of no interest to your client.

The investigation of family background, of the situation in which the client first experienced the challenges of life, provides the immediate context. This is why the *Summary of the Family Constellation* is completed before the early recollections are studied.

In the review an even narrower context is established for each of the recollections by considering the age of the client at the time of the reported incident. By attending to this we are alert to the special meaning of recollections associated with (for example) the birth of a younger sibling, a parent's dramatic illness or death, menarche, or some other event of consequence to the client.

2.   *Content.* It may not seem worth mentioning, but it is nonetheless important to note that the recollection belongs to the client. It is not legitimate to interpret a recollection based on what "it must have been like," or what "any child would have felt in that situation." It is one thing to practice empathy; it is another thing to impose one's own ideas upon a client's experience, or to assume the presence (though hidden) of material required by a pet theory. This is what keeps Freud's Psychoanalysis from being a reliable interpretive schema. By treating memories as "internal screens" (G., *Deckerinnerungen*), psychoanalysts are free to indulge in speculation about what "must be" behind the screens, shoring up the theory by postulating whatever the theory expects to be there.

On the other hand, it is equally important to pay attention to every detail included in the recollections. In fact, we could say that there is a law of inverse consequence here: The more extraneous a detail seems to the narrative, the more important it is likely to be to the narrator. For example, the color of an object does not alter the use of the object in a story. To mention color at all is to suggest a visual interest, perhaps an artistic one. If the color is vivid, or described in terms of temperature (cool blue; hot orange), it may imply a reference to subdued emotion, or to excitement.

3.   *Gender.* Some clients include only persons of the same sex among the characters appearing in their recollections. Some

reverse this, and tell stories with only members of the other sex present. Some (men and women) relate stories of good men and bad women. Others reverse this, and portray noble women suffering in a world of evil men.

No single issue is more important to social life than that of gender. To be clear about where we stand in our personal sense of gender, and to be at ease in our sense of gender among others, is central to our ability to develop community feeling, that liberating sense of being co-creators of the world in which we find ourselves. Whatever else the history of the past one hundred years may have taught us, we should be clear about this at least: No one, made to feel of diminished worth in comparison to others because of gender or gender identity, can be expected to cooperate cheerfully as a member of the human community. Moreover, no one made to feel that his or her value depends upon a position of gender superiority to others, can be expected to participate in community life without pretenses that threaten to disturb its peace, by insults, outbursts of rage, and other disguises of underlying insecurities and occasional panics.

4.   *Movement*. Life is movement. In a review of the early recollections, therefore, it is essential to see who moves, and how, in what direction, and with what effect. This gives the paradigm for the way in which the client is prepared to move, and expects others to move. We consider these five issues in examining movement:

i)   *Effectiveness/Ineffectiveness*:   What are the results of my actions? Are they positive or negative? Are men more effective than women? How? Vice versa? How? Are men and women effective in differing ways? How are these distinguished? Does suffering overwhelm power? Are others effective with me, while I am ineffective with them?

ii)   *Relative position*: Are some people above others? Does movement proceed from above to below? Is a vertical

dimension prominent in the memories? In certain of them?
What is the consequence, as seen in movement?

iii)   *Degree of activity and initiative*: Is the client physically
       active? Does the client initiate activities with others? Is
       movement limited to looking, listening, or resisting? Is the
       client passive while others act? A victim of others? Of luck,
       fate, or destiny? Or, a beneficiary of these, as in "a charmed
       life"?

iv)   *Extent of participation and cooperation*: Does the client act
       alone, or with others? Against others? Openly or in secret?
       As leader, follower, or in what role amongst others?

v)   *Sensory experience*: Is there one dominant mode? Visual?
       (Consider possible detachment, or artistic or design interest.)
       Auditory? (Consider possible interests in poetry or music.)
       Tactile? (Consider possible affect hunger and compensatory
       efforts to generate sensation.) Olfactory? (Although odors
       remain important to us, the sense of smell is greatly
       attenuated in human beings compared to its acuity in other
       mammals. If an odor is not pronounced we have to get close
       to the source for it to be apparent to us. References to odor
       may therefore be indexes to evaluations of closeness or
       avoidance, depending on the pleasurable-unpleasurable axis
       of feeling used to describe them. It follows that issues of
       intimacy can be indicated by memories of odor.) Taste?
       (Attention to pleasant tastes may reflect self-indulgence; be
       alert for other evidence that may suggest a sociopathic
       pattern.)

A final note on sensory detail: Someone may be inspired to
design research projects to test these clinical impressions. Until
results are available, use your intuitive sense to explore with the
client what a narrowing of attention to one or another sensory
modality may imply. Watch also for recollections with an aesthetic
continuum of sense data of every kind, sometimes reported as
"dizzying" or "transporting," and having about it an affect

described as intensely, almost intolerably, beautiful. Further inquiry may reveal this as a sign of religious disposition or of revolutionary ardor.

5. *Evaluation.* An inquiry into feelings has become a therapeutic cliché, so much so that a caricature pictures the therapist asking after nothing else, as if an empathic consequence would be the result, and that this were the only outcome required for successful therapy. By contrast, when we ask what the client's feelings were at the precise moment being recalled in an early recollection, we are after specific and important information.

Feelings are the index to evaluation, the key to an understanding of subjectivity, which is part of the proper subject matter of Individual Psychology. To say how I feel about something is to indicate where I stand with respect to it, how I am prepared to respond, with whom I am prepared to sympathize. For this reason some discouraged clients will be reluctant to let you know what they feel. To let anyone know where I stand, or even to take a stand of my own, is to put myself in a position that may be criticized or opposed by others. To escape this danger some clients will avoid responding to the question, "What did you feel at that moment (the most vivid moment recalled in the memory)?" Instead, they will tell you what they *thought*, and they will be happy to provide a long rationalization of their opinion. It is important to get beyond this if you can.

Sometimes the client cannot understand the question. There is some genuine confusion of meaning in the common usage of the phrases, "I think" and "I feel." For example, "I feel that the national debt is too large." In fact, there can be no such feeling. We should rather say, "I think that the national debt is too large." Of course we have no need to be so careful in everyday speech, since our meaning is generally understood by those who agree to use the words as loosely as we do. The client may therefore wonder why you are making an issue over the distinction. You must find a way to persist. The issue is clarity; the psychoclarity process depends on it.

Even so, you must not get into a quarrel about words. We sometimes get to the point by asking, "What were the sensations in your body?" That usually takes us a long enough step away from cerebration. The trouble with this tack is that (to give examples from client responses) "It's hard to define," or, "It's hard to put into words." Of course. We encourage: "That's O.K. Just describe the way you felt." If things stay blank we may go as far toward prompting as to say, "Well, did it feel good or bad?" Then, whichever of these is chosen, we can go on with, "What kind of good (or bad)?" If this fails, we try, "Well, was it on the warm side or on the cold side?" This can reveal whether the feelings were toward the conjunctive or disjunctive. Sometimes none of this works to elicit feelings. In such a case an absence can be regarded as a presence of another kind. A person may be unable to imagine how to take an action, or even make an assertion, without making matters worse than they already are. When this is so, all affect that might prompt an attempt is dampened. Also, one may limit a style as narrowly as possible to prevent attachment and the threat of loss, trying to treat all of life as a technical problem, or a spectacle, as if claiming, in playwright Christopher Isherwood's phrase, "I am a camera."

The absence of affect presents a danger to the novice in trying to make sense of another person's memories. The beginner may be tempted to assume what "any child" of that age, in that situation, "must" have felt. This is a grave error, which will leave you with nothing more than an interpretation of your own projections. One woman remembered fleeing her homeland on a boat crowded with refugees, hearing the sounds of artillery behind them in the hills above the water. "Any child" at four years of age "must" have felt terror and confusion. This woman remembered a tremendous pleasure, with so many members of her family close around her, the beauty of the sun on the water, and the excitement and sense of discovery at taking her first boat trip! But what about the artillery? "Well, I didn't mind that," she replied. "I knew that if it weren't for that we wouldn't have gone on the boat at all!"

Example 1: Shane

> I was about six years old and there was this gigantic snowfall. I
> went out into the driveway. I was supposed to shovel it.

> Most vivid moment: Standing there, seeing that huge drift.
>   How could it get that big? How was I
>   ever going to get out of that driveway?

> Feeling: Jesus! Futility. I was overwhelmed. It was too
>   much!

1.   *Context.* The client was six, and had started school. Life
outside the family presented itself, and presented the client with
responsibility.

2.   *Content.* There are two rhetorical questions: "How could it get
that big?" and "How was I ever going to get out of that driveway?"
The first of these conveys a sense of drama, wonder, excitement,
and an exaggeration of the difficulty; this is seen as so huge that
the client's problem-solving ability is swamped, and the answer to
the second is implied: "I'm not. There's no way that I can solve
this."

3.   *Gender.* Since the client is a man, the recollection may be
stating, "Look at what a man is up against in this world." There
may be a further implication: "Since it looks as if it's too much for
me, I must be an inadequate man." Note, too, that he is all alone in
facing life's difficulties. That may be the man's lot in life.

4.   *Movement.*

i)   Effectiveness/Ineffectiveness: I am ineffective.

ii)   Relative position: I am small and the world and its
      difficulties are "huge."

iii)   Degree of activity and initiative: Since I am in a world filled
       with insurmountable obstacles from which there is no escape
       ("How was I ever going to get out of that driveway?"), there
       is nothing I can do.

iv)   Extent of participation and cooperation: A man is on his own.
       There is no aid; there are no allies. I am one man, alone in a
       cold world.

v)   Sensory experience: The emphasis is on the visual. I am
       impressed by what I see, and what I see is bleak, colorless,
       immense.

5. *Evaluation.* Life's struggles are as futile as they are inescapable.

Example 2: Doug

> I was about three. My mother and I were at the beach. I was
> playing in the sand. There were lots of birds flying out over the
> water. Suddenly my mother grabbed me up and took me home.
>
> Most vivid moment: Her scooping me up.
>
> Feeling: Mad.

1. *Context.* The client's younger brother was born at about this
time, and Mother was no longer as free to indulge him. The
contrast between his former (and still claimed) freedom, and his
being required ("scooped up") to yield to her wishes shows itself
here.

2. *Content.* Most striking here is the statement, "There were lots
of birds flying out over the water." This detail, extraneous to the
drama of the story, succeeds in providing an imagery by which to
highlight the client's "antithetical mode of apperception" (Adler,
p. 248). It contrasts the absolute freedom of the birds, and the
constraint he experiences at the hands of a woman (see below).
When we asked him about the birds he said, "I always wished I

could fly like a bird." We asked him what that would mean to him and he replied, "Freedom."

3. *Gender.* In this story it appears that a man is at the mercy of the woman in his life, and she decides what is required of him. His interests don't seem to weigh anything in comparison to hers, and her determination and strength are irresistible.

4. *Movement.*

i)   Effectiveness/Ineffectiveness: I can be effective when I am working on my own projects, and I am left alone. I am ineffective against the determination of the woman in my life.

ii)   Relative position: I am no match for the woman in my life; no matter how disposed I am to fight back.

iii)   Degree of activity and initiative: I can take initiative on my own, when left to myself. Against the woman in my life, I can take no initiative. She has her way with me, and my anger is of no consequence. She is in charge.

iv)   Extent of participation and cooperation: I can either be on my own, and able to enjoy what I want, or I can have a woman in my life to whom I must submit. There is no cooperation in either case.

v)   Sensory experience: Visual (birds); tactile (playing in the sand).

5. *Evaluation.* I am in an impotent rage, trying to impress the woman in my life with my desire to have some freedom, without having to give up her caring for me.

<u>Example 3: Adam</u>

I was about seven. My big brother and I and a bunch of boys were playing basketball in the back of the house. My brother was bigger and taller than I and kept blocking me so that I couldn't get the ball.

Most vivid moment: My brother with his back to me, keeping me from getting the ball he was dribbling.

Feeling: Frustrated. Angry.

1. *Context.* This is the memory of one of the younger boys in a large family. Here he is at age seven in the boys' world, with his older sisters not around to protect him, and with his big brother keeping him in his place.

2. *Content.* We are struck by the image of the brother's back, turned as if to underscore a feeling, familiar to many younger children, of being disregarded, not taken seriously, or insulted.

3. *Gender.* This is the way it is in the man's world. It's a rough game, a competitive game, and it's hard enough to get into it, much less to get ahead in it, if you're not as big as the others.

4. *Movement.*

i)   Effectiveness/Ineffectiveness: I'm not a big enough player to make a success of things. The bigger players block me, and keep me back.

ii) Relative position: I'm coming from behind, looking for a break.

iii) Degree of activity and initiative: I'm game. I keep trying. I'm full of fight.

iv)   Extent of participation and cooperation: I'm right in there among the other guys. I'm playing by the rules, but I'm watching for my chance, and I don't expect anyone to hand it to me.

v)   Sensory experience: Kinesthetic and tactile. The rough and tumble.

5. *Evaluation.* I am up against unfair odds. I am determined to hold my own and to fight for what I want.

<u>Example 4: Denise</u>

I was five. It was my birthday. The party was in my grandfather's back yard. There were about ten children there, both boys and girls. I remember the present my grandfather gave to me, because it was just what I wanted: a buggy for my doll. It was the most beautiful thing I had ever seen. Just as I was running over to get it, one of the boys fell into the fish pond. It was about nine feet deep. They pulled him out and he was covered with tadpoles. They were in his hair, on his clothes.

Most vivid moment: There were really two: The first was when grandfather brought out the doll buggy.

Feeling: Overwhelmed with joy!

Second most vivid moment: Seeing that boy covered with tadpoles.

Feeling: Amused. It was hilarious!

1. *Context.* This client's other recollections showed women as competent and caring. Now we have a contrasting memory telling us about men. This is presented to us at a time in the client's life when she was experiencing marital difficulties, and complained

that her husband was "insensitive to my needs." (He complained that she criticized him and put him down.)

2. *Content.* Two pictures of men are produced, as if simultaneously, showing (1) that a man can be complete and mature, can know just what it takes to make life beautiful for a woman, and can be counted on to do it without having to be told; (2) that other men are inept, get in over their heads and have to be rescued, and appear ridiculous in their incompetence and immaturity (among the other tadpoles).

3. *Gender.* Men either provide for or amuse me. As a woman, I can be delighted with what they provide or entertained by what they do.

4. *Movement.*

i)   Effectiveness/Ineffectiveness: As the star, I am a centerpiece for the men to fuss over, provide for, or amuse.

ii)   Relative position: I am at the center of it all, in the happy position of having a worthy man demonstrate that he can please me by giving me "just what I want(ed)"; and of having a raw, unformed man demonstrate that he is not yet equal to anything more than amusing me.

iii) Degree of activity and initiative: Others act; I evaluate, and I experience either overwhelming joy or hilarious amusement. Nothing touches me to hurt me.

iv)   Extent of participation and cooperation: I am at the center of the fun as the star of the show. If it weren't for me the fun wouldn't be taking place. I enjoy what the others do.

v)   Sensory experience: Visual. The recollection is cinematic.

5. *Evaluation.* The "overwhelming joy" and the "hilarious" extent of her being "amused" show her as seeing life as impinging on her

in a shower of blessings, for which she is qualified simply by being alive (the celebration of her birthday). Requirements rest on the others, and only those who know what they are doing can fulfill them.

An Illustrative Case

Now that we have said so much about the elements of interpretation, and about the categories into which we attempt to sift the data, we offer a complete transcript of a session on early recollections as an example of this work. If you find it difficult to compare our interpretations point-by-point with the preceding categories, this is as it should be. When it comes to the living work of understanding another human being, these "key categories" are not sovereign; the person is. They shape our thinking, but the work itself shapes the outcome.

The transcript is of the review and interpretations of the early recollections in "The Case of Janice." To establish context, we have included the summary of her family constellation so that you will know what we knew as we entered this interview with Janice. Bear in mind as you read the summary of family material and the subsequent transcript that Janice's presenting complaint was acute anxiety with panic reactions relating to concerns about her twelve-year-old son, a well-behaved, cheerful youngster on the threshold of adolescence, who had never been a problem child, in any sense of the phrase. Janice was preoccupied and worried about his activities, and the possible harm that might come to him, to the point of incapacitating herself in a flood of symptoms.

The Case of Janice

Summary of the Family Constellation

Janice grew up as the older of two children and the only girl, psychologically the firstborn daughter, in a tradition-bound New England family in which the atmosphere was constraining, as if each of the parents was concerned to hold the energies of the children in check by managing the expression of those energies and channeling them into athletics and family-centered activities.

The family values emphasized self-control, propriety, and privacy (if there was trouble, "no one should know about it"), which alerted the children to the idea that there was little reason to repose confidence in others, or in life.

The masculine guiding line set by Father stressed hard work, caution, and an uneasy pessimism which expressed itself in faultfinding or, when no fault was to be found, in fault-seeking. Father was the unhappy firstborn son of his own dominating, cold, and withdrawn father (who confirmed the character of the masculine guiding line for Janice).

The feminine guiding line set by Mother stressed gregariousness, "ladylike" composure (Mother kept her own counsel, and never raised her voice), and "feminine" interests in clothes and appearance. Mother was also, however, athletic and robust, and rejected her own mother's rejection of men.

Ralph, four years younger, made a place for himself as the firstborn son who imitated Father in a quiet determination to have his own way, in being self-absorbed, and in being critical of the women.

Janice kept her place by being a better student and by doing her best to meet the parental demand for decorum and restraint.

During childhood she chose her father as a role model in terms of withdrawal, isolating herself in the world of books, and consoling herself against the constraints of her situation by picturing the other places, other times, and the lives of other people that her books disclosed to her.

In adolescence her success in following her chosen masculine role model became problematic. As a girl who would be a woman, she aspired to enjoying advantages inhering in the kind of position Mother enjoyed: that of a pretty, soft woman, fussed over, protected, and maintained by a powerful man (like the nineteenth century heroines she admired in novels). She succeeded in being like Mother in terms of gregariousness and enjoyment, but never came to see herself as being sufficiently feminine. On the one hand, she fell short by the measure of being soft-spoken and by the measure of the suppression of anger and resentment; on the other, she deviated through her determination to make an independent life for herself (though not without taking note of the fact that Mother had also arranged to do as she pleased, as long as Father didn't know about it).

Prominent among other images of femininity available to Janice was that of her Aunt Sarah, who was "trapped" in an unhappy marriage to a harsh and unfriendly man.

Through everything in Janice's childhood and adolescence runs a scarlet thread connecting a variety of ideas about control, self-control, domination, and resistance — issues expressed in all the family relationships. Her pattern of finding excitement in numerous "sneaky" ways rehearsed a longing to escape from constraints, to take risks, and to experience the liveliness of life. Even here, however, the effort was secret and muted by her sense that, above all, she had to "stay in control."

At the threshold of adulthood, the question remained: How could she, a woman, stay in control in a world dominated by the men?

REVIEW AND INTERPRETATIONS OF
THE EARLY RECOLLECTIONS

[Note: The record of Janice's eight early recollections appears here. Each is followed by a verbatim transcript of the discussion that took place among Janice, the recording therapist (JG), and the consulting therapist (RLP).]

ER 1: Age 3. We went on a trip to our summer place in Maine. We had an MG roadster with white upholstery. You could fit in there behind the seats with your pillow. I'm in there. It's night. My parents are talking.

> Most vivid moment: Being in there, hearing my parents talk.
> Feeling: Safe, good, warm, happy.

Discussion

RLP: It's an interesting first memory, visually. An MG with white upholstery. And I'm tucked away. An MG with white upholstery sounds pretty flashy to me.

Janice: It was hysterical. My mother was able to make Dad give it up after my brother was born. My father could get the flashiest car he wanted before he had children, and it was all right with Mother then. (All laugh.)

RLP: And the feeling is secure?

JG: Safe, good, warm, happy. She's tucked in there, looking out at everything, and listening.

RLP: Were you looking out?

Janice: Yes. You two are a trip! (Laughter.)

RLP: What are you thinking?

Janice: Nothing. (Laughter.) It's just fascinating. You two have never been on this side.

RLP: Oh, yes, we have! (All laugh.) There's a sense of having a place of my own, among others. They're doing their thing, and I'm doing mine. I'm in the world of the others, along for the ride, but I have my own place. O.K.

[Note that Janice's participation in the discussion of her first memory is an illustration of the meaning of the memory, and a demonstration of her style of movement in distancing herself from others, here disguised as humor. Even her choice of slang ("You two are a trip!"), and her fictional location of herself on a "side" of the transaction, recapitulate the situation of the memory. We could probably go further, and say that her laughter over her father's penchant for flashy cars should have served as a warning to the therapists to the effect that she was prepared to ridicule ("It was hysterical") and thereby diminish any way, not her own, of looking at things. This is a parade example of what Adler called the "depreciation tendency" (Griffith & Powers, 2007, p. 23).]

ER 2: Age 7 or 8. My brother had a snake in a tin can in the kitchen on the floor. In the dark I got up for something. I kicked the can over, then I stepped on the snake. Then I remembered the can was there.

Most vivid moment: When I felt that thing under my foot.

Feeling: Fear.

Discussion

RLP: Then I remembered! After I kicked the can, after I stepped on the snake, then I remembered.

JG: After stepping on the snake. It's certainly a cautionary tale.

RLP: Except it says what I should have done. Events go out of control. I knock things out of control. I should have known about it.

Janice: Good time to remember the can, huh? (All laugh.)

JG: There's an interesting contrast between the first and second memories. In number one, she's going along for the ride, and other people are in charge; in number two, she's in charge — that is, acting on her own — and she blunders.

RLP: Yes, here she's taking her own initiative and it doesn't work out. O.K.

ER 3: Age 8 or 9. The house across the street was very old. They had a cellar that had a mud bottom. There was a storm door. There's water collected in a pool in that basement.

Most vivid moment: I'm standing there in that basement with my girlfriend, staring at the pool.

Feeling: Good, peaceful.

Discussion

RLP: Nothing moves. Just still water.

Janice: Yes.

RLP: What was so good and peaceful?

Janice: Water has always done that for me.

RLP: Tell us what you mean, what comes to mind when you say that.

Janice: Warm, contained . . . Umm . . .

RLP: That's all right. You're doing fine.

Janice: I mean it was just so peaceful in that basement.

RLP: There's something about the contained character of it in that pool.

Janice: I don't necessarily . . . I was just thinking about it . . . I don't necessarily like things that are endless, you know? I mean when I look out at the ocean . . . And you just look out . . . And you can't see the other end. I don't necessarily like that.

RLP: All right, we've got three . . .

Janice: It was dark. It was dark down there, too.

RLP: . . . we've got three images of containment and uncontainment now.

JG: And the dark.

Janice: Um-hum. I was always more comfortable in the dark.

JG: In the car, in the kitchen, in the basement.

Janice: Um-hum.

RLP: We've got containment in all three of them, though the snake is uncontained in the second memory.

JG: And she's fearful when it's uncontained.

RLP: Yes. It's the "life force," run amok. Things that are not contained, energies that are not confined. It's scary to you.

Janice: Um-hum.

RLP: O.K.

ER 4: Age 4 or 5. I went to buy shoes at the department store and to see Santa Claus.

Most vivid moment: Sitting on Santa's lap.

Feeling: I was not real comfortable. I was shy.

Discussion

RLP: What was uncomfortable about it?

Janice: My mother wanted this picture with Santa Claus in it . . .

RLP: Yeah?

Janice: I didn't necessarily want to do it, I don't think.

RLP: See if you can tell us what you remember about it.

Janice: I didn't know him.

RLP: Yeah?

Janice: I mean I'm not real good at . . . I don't think I was as a child either, real good at things that I didn't know. It took me a while to get used to things.

RLP: Um-hum.

Janice: And I think a lot of times, when I felt that way initially, I never gave myself the time to get comfortable. Once I get comfortable, I'm fine. It took me a lot of years to discover that, I think. To be able to immediately feel O.K.

RLP: O.K.

ER 5: Age 6 or 7. It's late. My mother and father were in their bedroom. I woke up — maybe I had a dream, I don't remember — and went in to them. They were having intercourse. I thought they were hurting each other. There were two white bodies, like marble.

> Most vivid moment: Seeing the whiteness of the two
> bodies and motion. Father was on
> top, and Mother was on the bottom.

> Feeling: I was frightened.

Discussion

RLP: Did they know you were there?

Janice: I never got up again, I'll tell you! (All laugh.) Yeah. They did, because I . . . I went in anyway.

RLP: What happened?

Janice: I think they put me back to bed. They must have. They didn't get real bent out of shape.

RLP: What happened? What are you remembering?

Janice: I think . . . I think they just stopped and pulled up the sheets and took me in bed until I fell asleep and then put me back in my own room.

RLP: They took you into their bed?

Janice: Um-hum.

RLP: So you . . .

Janice: Yeah. I would say they did.

RLP: So your interpretation was that they were hurting each other? Or that he was hurting her?

Janice: Yeah. Yes. The latter.

RLP: That he was hurting her.

Janice: Yes.

RLP: What did you make of it?

Janice: God knows!

RLP: You're six or seven years old. What do you know about intercourse? All you see are these two people.

Janice: That's right, and hear the . . . the sounds bothered me.

RLP: What sounds? What was there about the sounds?

Janice: I don't know . . . uh . . . uninhibited, I guess, not things I'd heard before, nothing I could relate to.

RLP: The idea you form is that he was doing harm to her.

Janice: Yeah.

RLP: As if what? Since you didn't know what intercourse was, what did you think he was doing?

Janice: I don't know . . . You know, like somebody hurting somebody, like when you see people pinned down.

RLP: A struggle?

Janice: Yeah. Yes. Yeah. Not pleasurable . . . I didn't think of it as that at all.

RLP: Somehow uninhibited, unrestricted . . .

Janice: Yes, but that he was in charge.

RLP: Overwhelming. It looks like a dangerous struggle.

Janice: Yes.

RLP: O.K.

ER 6: Age 8. I'm dressed up for Thanksgiving. I'm out in the front yard of a good friend's house, waiting for my parents to get ready to go to my grandmother's house.

Most vivid moment: Being out on the lawn, having my picture taken.

Feeling: Good.

Discussion

RLP: Who's on the lawn?

Janice: Me and my friend.

RLP: Same age? Two kids, huh?

Janice: Yeah.

RLP: And who's taking your picture?

Janice: My father.

RLP: Help me to see where you are in the memory. He's taking a picture of the two of you?

Janice: Um-hum.

RLP: Where is he?

Janice: We're together — Judy and I are together. And he's back there.

RLP: Back where?

Janice: Father's over this way, taking our picture. We're here, and he's there [indicates]. He's taking the picture from there. I have a red hat on.

RLP: Um-hum. You have a red hat on?

Janice: Yeah. And a new wool coat. We were all excited!

RLP: Look at the situation: Here's a man at a distance and the women all dressed up. The men act as if they're interested in how the women look, and the women make the pretense of looking good. But the truth of life — from the prior memory — is that the men will overwhelm them.

Janice: Um. Yeah.

RLP: What was the excitement about that day?

Janice: Just Thanksgiving. You know, going to Grandmother's house and all that business. Thanksgiving in our family was always a big deal.

RLP: Um-hum. O.K.

ER 7: Age 9. There were some people who were friends of the family. They had a Halloween Party in their basement. I went to the party as a ghost. I remember bobbing for apples. I had a great time.

Most vivid moment: Bobbing for apples.

Feeling: Good. I liked it. I didn't catch any, but I liked it anyway.

Discussion

RLP: What was there about it that you liked?

Janice: I want to say, just listening to you say this, that it was because I was the ghost, and I didn't care what anybody thought because nobody could see who I was. It didn't matter whether I caught the apple or not. But I don't know why I want to say that. I really don't.

RLP: Well, is it true, or are you making an adult interpretation? Are you remembering that, or are you speculating about it on the basis of present assumptions that . . .

Janice: I think that's true, that I was covered up in my sheet and I enjoyed that. And it was . . . God! It was dark in that basement! (All laugh.) I don't want to say that word, dark, any more! But . . . I liked scary things . . . I still do. We took a trip on city buses last week to visit some relatives. (Laughter.) I loved it! My kids thought I was crazy. I mean, we'd never taken the bus there before, so I didn't know where I was going. I mean, we got lost! We got off at the wrong stop and stuff, and I thought, "This is so exciting!" And my son said, "Mom, you're really enjoying this." And I said, "I love it! I've never done this before!" And I thought of you two then, and what you'd said about excitement, and I thought, "It's true. I really do like excitement."

RLP: O.K.

ER 8: Age 9. We're climbing trees out in the back of the house next door to ours. The really big deal was to jump from the tree onto the roof. This was the house where two of my brother's friends lived. We climbed the tree, jumped onto the garage roof, and, from there, onto the roof of the house.

Most vivid moment: Hiding on the other side of the roof from the door to the house. Ted's mom was coming out of the house to hang out the laundry. These were pitched roofs, and we kids were holding on to one side while Ted's mom was coming out of the other. We're hoping she won't see us.

Feeling: Good. I was thinking, "I still didn't get caught!" It was an endurance contest: You didn't know how long you could hold on! Or, when you might fall, and be discovered.

Discussion

RLP: Getting caught . . . doing what? Taking part in a boy's game? Hiding up above, where the boys are — the men are, up above the women? Do you have some feeling in your life that you're "hanging on"?

Janice: I don't know. I was just sitting here thinking about that.

RLP: How long can I hang on? Do you have some idea about that?

Janice: I feel like I'm hanging on to get through something.

RLP: Like what?

Janice: I had that feeling when I was in college.

RLP: Do you have it now?

Janice: Not now, but when I first came in here to see you two I had that feeling: "If I don't get myself some place, I won't be able to hang on much longer." I was so anxious then.

RLP: O.K. Let's review the entire record now. I want to run through everything again with ages and situations.

JG: O.K.

[Note that the discussion now continues in the "second review" of the recollections, to which we referred on p. 279.]

RLP: Now, the first one is in the back of that car. How old was the child?

JG: Three.

RLP: Feeling?

JG: Safe, good, warm, happy.

RLP: All tucked away.

Janice: Um-hum.

RLP: And other people are in charge. Is the car being driven? Is it in motion?

Janice: Um-hum.

RLP: So it's in motion. So I'm being carried along. There's some sense of being safe and sound and tucked away and letting somebody else have charge of life. Let's put it that way. And it's O.K., it's fine. I'm safe where I am, sort of hidden away.

JG: And the next memory is at age seven or eight with the snake in the kitchen.

RLP: The first was age three, and then we jump up to seven or eight. She tips over the can and then realizes that there's a snake in it and there's a sense of fear.

Janice: Um-hum!

RLP: And then she steps on the snake. Are you barefooted?

Janice: Yeah!

RLP: And here's this wriggling life under my foot, eh?

Janice: Um-hum.

RLP: It's hard to keep it there.

Janice: Yeah!

JG: The next one's at age eight or nine in the basement, looking at the pool, and it's peaceful.

RLP: Eight or nine. Now as I recall, your first period came when you were nine, didn't it?

Janice: Um-hum.

RLP: There's something about the containment, the stillness of the water. O.K.

JG: The next one is sitting on Santa's lap, back at the age of four or five.

RLP: We go back in age now, to the time when your brother was born, and here I am on Santa's lap. I don't know who he is, but I'm supposed to be sitting here. I'm supposed to be enjoying this?

Janice: Um-hum.

RLP: I don't understand the intrusion of this memory into the record at this point in the sequence. Let me get it again: We have the car, the snake, the water — all in the dark. Then we suddenly emerge into the light, and here I have to sit on Santa's lap. I don't like it, right?

Janice: Um-hum.

RLP: Who is this person? What am I supposed to do here? What was the feeling you had?

JG: "Not real comfortable; shy."

RLP: It indicates some reluctance to engage with him. It's really the first time there's any kind of social . . .

JG: No. There's the other girl at the pool. . . .

RLP: Yes, but she and I just stand there, and look at the pool of water. The focus is on the water, isn't it?

JG: Yes. "I'm standing there with my girlfriend, staring at the pool."

RLP: Well. Then we come to the first "contact" memory, and it's with a man. We can look at it that way: A man (Santa) and a struggle. All right. What's the next one?

JG: Mother and Father.

RLP: Well, then it does connect. It's a shift there at that point. There I am made to sit on a man's lap. Here I witness this man overwhelming this woman. It's a scene that's overwhelming. It's in the dark. It's the hidden reality between men and women . . . as if they were a tableau. As if they were a sculpture. It's as if I were seeing something like the secret of what life is all about. And it's not very pleasant: What it's all about is that men overwhelm women, in some strange way that really isn't very clear, but with signs (the sounds) suggesting that the woman acquiesces by abandoning her inhibitions.

JG: The next one is the Thanksgiving picture on the lawn with the girlfriend.

RLP: Now we get back into the company of the girlfriend immediately, and Father, the man, is at a distance, taking the picture. There's a great deal of excitement, pleasure.

JG: In the company of the girlfriend, the feelings are good in both memories; and they're standing still — just looking or being looked at. Nothing happens.

RLP: You're in a red hat. The Freudians would love that. You know how they get all lashed up over Red Riding Hood and the menarche. You're in literary publishing; you read *The New York Review of Books*.

Janice: Right. I've read more than enough of that stuff!

RLP: O.K. And the next one?

JG: The next one is the Halloween party as the ghost.

RLP: Just think how much fun the analysts would have with this — and how far away it would take us if we started doing all kinds of symbolic things with menstruation, and sheets, and bobbing apples!

JG: Well, then, let's just stay Adlerian, and we could say it's all right to fail. I didn't catch the apple, but I liked it anyway. I liked the game. It didn't matter — to win or lose didn't matter.

RLP: That issue suddenly shows up. I see. Well, winning or losing doesn't matter as long as no one knows who I am. O.K.

Janice: You two guys are something else! (Laughter.)

JG: Well, thanks. (All laugh.) The last one is climbing the tree, and hiding on the other side of the roof.

RLP: Um-hum. Out of sight.

JG: Yes.

RLP: Who was with you in this climbing memory?

Janice: My brother's friends, Ted and John.

RLP: And you're all three hanging on the other side of the roof, and it's Ted's mother who comes out to hang wash?

Janice: Yeah. (Laughter.)

RLP: There're three of you hanging there, being still, so she won't hear you?

Janice: Right.

RLP: Where are you among the three?

Janice: In between Ted and John.

RLP: And the feeling is, what?

Janice: Excitement. Yeah.

JG: And "good." You still don't get caught. "It was an endurance contest." You didn't know how long you could hold on.

RLP: What is the sense of this memory? In the memory you just stay there, hanging on. I'm enduring to see how long I can hang on — until what? Either I hang on until then, or what?

Janice: Either she goes in, or . . .

RLP: Either she goes in, or I lose my grip?

Janice: Yeah. And then, if you fell in the cinders down below, she'd hear you, you know.

RLP: If you'd slide down the roof, there'd be cinders on that side, huh?

Janice: Yeah.

RLP: You'd be giving yourself away?

Janice: Um-hum. You're praying she'll go in so you don't get caught . . . 'cause you can only hold on so long. And then, she can hear you, you know, your shoes? You know how that old tarpaper was?

RLP: Yeah.

JG: So there's nothing about if I fall I'll get hurt.

Janice: I know I will. I know I'll get hurt.

RLP: But apparently that's not the point, or at least not the worst of it. The point is that you'll "get caught." By her. A woman will see something about you that you don't want her to see.

JG: These are boys, and there's camaraderie, closeness. She's in it together with them.

RLP: Where she shouldn't be. That's the end of the record?

JG: Yes.

RLP: See, the record starts and ends with a kind of hiddenness, a being hidden away. It ends with a different kind of hiddenness from that with which it begins. Was this in the light of day, this last memory?

Janice: Yeah. Um-hum.

RLP: You're still out of sight. How do you know she's still in the yard?

Janice: We see her come out the door.

RLP: How do you know how long she'll be in the yard?

Janice: We don't!

RLP: How will you know when she's gone back in?

Janice: We'll see her go in.

RLP: How will you see her?

Janice: Over the roof.

RLP: Hanging there?

Janice: Yeah. It's like you're hanging like this [indicates].

RLP: Oh. So you're peeking. So you can see her across the top of the roof, but she can't see you.

Janice: You've got it.

RLP: Well, we're on to something important, but I still don't feel that I've "got it" in the sense that I can understand it. There is something about making excitement in your life in some secret way. What is your way of stirring up excitement? And what is your secrecy about? People for whom control is a big issue frequently do it by stirring up emotional storms. It's a form of internal excitement that can't be seen by others.

Janice: What do you mean?

RLP: Some kind of "Whoops!" feeling, something beneath a calm exterior. I don't know how you do it. What are your thoughts about it? I don't get this very clearly at *all*. I don't know why I say, "not *very* clearly"!

JG: Since "hiddenness" is one of the themes here, that could explain why they aren't very clear.

RLP: Yeah. The memories are doing the thing they're about. Good point. Let's see: I'm hidden away, carried along. In the very first one, if we were to say it this way, "I'm hidden away, carried along in the stream of events, interested in what's going on . . ."

JG: She knows what the others are doing: She's looking around, and listening to the parents talk. They don't know what *I'm* doing. And she watches things. "I know more than they know; I'm in on their secrets, but they're not in on mine."

RLP: It's as if I'm a hidden observer, huh? That's one dimension. There's also the dimension of coziness, of safety, behind the seats.

JG: And even safety on the roof. She believes she won't get hurt, and the two boys are with her, one on either side.

RLP: But, there's something there about being careful. In the first one, there's not the danger of being caught; they know I'm back here. It's just that I can listen to them talking. I'm taking in what goes on around me, but no one can see *me*. I'm all tucked away in this little hidey-hole. And the car is in motion: I'm "carried along."

JG: In the second one, about the snake, it seems that she can blunder into something. And then, look what happens if I'm not careful.

RLP: Well, that's one place where she blunders, kicking the can over and stepping on the snake. Then there's the other memory, of blundering upon her parents.

JG: So, "in the dark" may be the common metaphor for not knowing what's going on.

RLP: Yes, at least not knowing enough to act on my own, to be one of the people who makes things happen. In the first memory, I'm all cozy on the pillow, and I'm not acting. The others are in charge. Now the third one is the stillness of the water in the cellar under the house. Just observant, just contemplating, and feeling peaceful.

JG: Both of the memories with the girlfriend are still.

RLP: Water is a classic symbol in mythology, in dreams, and so on. If we were Jungians, what would we make of this?

JG: Water would be the womb.

RLP: And so, by extension, the female principle.

JG: And the snake, the male principle.

RLP: I wonder if we shouldn't take it seriously.

JG: Especially since they're juxtaposed: There's the fear of the snake and the peaceful feeling of the water.

RLP: The male principle here is wild and uncontrollable, and the female principle is serene and bounded. Let's assume there's something to it. What's the use of studying psychology if we don't talk about symbols? (All laugh.) What comes after the water?

JG: Santa Claus.

RLP: Ha! There's the masculine principle embodied in a mythic figure of power and bounty, and I just sit there. I don't know what to do in the face of it.

JG: She just sits there. She . . .

RLP: What does *he* do?

Janice: Nothing. What *do* they do? They ask you what you want for Christmas.

RLP: How does he act?

Janice: Not any different than with the other kids.

RLP: Let's be in the memory. How old are you? Three or four? Still a very tiny child.

Janice: In kindergarten.

RLP: A kindergarten child. Now, see through the eyes of the child. See if you can get the memory back. Mother stands back to take a picture. You sit on this strange man's lap.

Janice: I didn't want to go in the first place.

RLP: What are you remembering?

Janice: I usually . . . this was a department store near where we lived. You walked up the stairs and it was the shoe department. And at Christmastime, Santa was in one part of the shoe department. I liked that department store, but I did not want to go and see Santa Claus.

RLP: All right. Now, what happens? What do you remember?

Janice: Hanging back, dragging back, while she walked me up there. I knew I was going to have to go, whether I wanted to or not.

RLP: Yes?

Janice: And then I felt yucky when I had to get up on his lap.

RLP: How did you do it? How'd you get up there?

Janice: He had to help me.

RLP: How'd he do it?

Janice: He picked me up.

RLP: What was "yucky" about it when he picked you up?

Janice: Oh, I just squirmed to get away! I remember that!

RLP: Tell me everything you see as you squirm. What do you remember?

Janice: He was stronger. I mean, once he'd lifted me up, I landed on his knee, so I couldn't get away.

RLP: And?

Janice: I didn't cry or anything. I knew it wouldn't have done any good. She wanted the picture. (Eyes filling with tears.)

RLP: That's O.K. Just let it come forward. Stay in the memory. So? You land on his knee. You know you can't get away. He holds you there. He's stronger. It doesn't make any difference: There's no need to struggle . . . What are you remembering?

Janice: That once I relaxed, I felt O.K.

RLP: It's yielding. It's the yielding. And then, some sense of relief.

Janice: Yes. That's how I've felt. I would say that's the basic part of how I relate to most men. (Tears.)

RLP: Can you tell us about that?

Janice: I don't want to.

RLP: What comes to mind instead?

Janice: You know I was thinking about that back when I was telling you about my mother originally, back a few sessions ago, when you were recording all this stuff. I didn't like seeing him (Father) in the dominant position. That frightened me, I can remember. But I never feel comfortable when I'm involved sexually myself any place but underneath. You know. What do they call that position?

JG: The missionary position.

Janice: Yeah.

JG: Well, that's a hidden position. Other ways, you're exposed.

Janice: (Laughter.) Yeah. I was thinking that. I'm getting into this too much! (Laughter.) I was thinking . . .

RLP: Well, how much can you get into it? It's your life.

Janice: Yes, I know. But I don't even want to think about it. Anyway . . . I'm like that. That's a big thing for me. It took a lot of years before I was willing to play around with that at all. Like, I'm vulnerable if I'm on top, you know?

RLP: To what?

Janice: I don't know. To getting hurt. To not being in control.

RLP: How are you in control when you're supine?

Janice: (Laughter.) I don't know. I can't get hurt as much. That's how I feel. I can just curl up and not get hurt as much.

RLP: Just submit, eh?

Janice: Yeah, and not be as actively involved, so then, if it turns out bad, I won't be as involved in the first place, you know what I mean? (Laughter.) I mean, it wasn't my responsibility, because I never got involved at all. Yeah. I'm not responsible.

RLP: No, you're not. You're only enduring. As with Santa Claus. The men subdue the women. When Santa holds you, you don't act . . . What was that verb you liked?

JG: Yield.

RLP: Yes. You don't act; you *yield*. That's the whole thing: The man acts, the woman yields.

Janice: Yes. Somehow that's the whole thing. The man acts, the woman yields.

RLP: And then there's a release of tension. Like "falling off the roof." Do you know what that means?

Janice: Falling off the roof? (Laughs.)

RLP: It's old-fashioned slang. A sort of secret code among the girls, in the old days.

Janice: Yeah. (Long laughter.) It's your period! (More laughter.)

RLP: (With the laughter.) That's right. Give *some* credit to the Freudians! Their fascination with Red Riding Hood isn't *all* misplaced! Your first period came when you were about nine years old. (Quietly.) That's how old you are in the story of being on the roof. Hanging on, among the boys, wondering how much longer you can get away with it. That's the feeling you had when you were in college, and that's the feeling you had when you first came to see us. That's what you said.

Janice: (Now very serious, and calm.) That's true.

RLP: And we know that when you were in college at UMass [University of Massachusetts] you had an experience that was hard for you. You fell in love, and defied your religious training by active sexual involvement with the man. More, you defied the standards of your community by publicly choosing a man of another religion and nationality. When you became pregnant he abandoned you, and you had to accept all the responsibility of the affair without anyone to help you. You gave the child up for adoption, and you have felt the loss of that child ever since. Is all that a fair summary of what you have told us?

Janice: Yes. It's all true.

RLP: And you swore that you'd never let yourself be put into that kind of position again. And you've played a much more careful part, making your husband "responsible" for what happens, in the privacy of your bed, and in the "big decisions" of your life, while at the same time pursuing your profession, and attaining a position of some eminence in publishing without letting it intrude upon the privacy of your family life. Right?

Janice: You're right on. It's amazing.

RLP: It's only what you have told us. And now, the crisis that brings you to see us is connected to your son (the oldest "known," but not the oldest of your children) who has just turned twelve years old.

Janice: (Crying.) I don't get it. What's the connection?

RLP: What happens to boys when they turn twelve?

Janice: What do you mean? (Laughs.)

RLP: Come on. (Both laugh.) You know better than that! I'll give you a hint: What happens to boys at twelve that happened to you when you were nine?

Janice: (Laughter.) Oh, gosh. You want me to say, "puberty"?

RLP: I don't want you to say anything. I just want you to see what you have allowed us to see. Until now you could "protect" your son from all harm, or at least console yourself with the idea that you might do so, that it wouldn't be beyond you. You could even share the exuberance of his boyhood with him: Your adventure on the buses last week is an example.

Janice: Yeah. That was a blast.

RLP: Exactly. It's exciting to play at staying in childhood, prolonging the carefree fun of it, seeing how long you can "hang on" to it with him. But the game is up. He's becoming a man, and you can't control the energies that are coming up in him now, and that will take him away from you sooner rather than later — at least if all goes well!

Janice: (Crying.) I know. And I don't want to stop him. I know that's nuts.

RLP: Exactly. That's nuts. To want to keep a growing boy underfoot! That's why you experienced your anxiety the way you did. Life was insisting that you go forward with your son's maturity, and you didn't want to go. You expected it to ask more of you than you could handle. How could you keep up your fiction of passive little woman versus active big man, and still keep an eye on your growing son, and try to control his activities?

Janice: A tall order, huh?

RLP: And how. But an even taller one, a much harder task, was to find a way to maintain your protective, controlling management of your little boy when you knew that was nuts, as you put it. That's why you couldn't put it into words. You knew that your anxiety had something to do with your son; you knew that it was somehow connected to your college experience, or, as you put it, to your guilt; and you knew that it was somehow all snarled up in your private notions about masculine and feminine. All you could feel was the embarrassment of it, as if you, a mere woman were trying to control your son, a more and more powerful man.

Janice: Whew! You two guys are too much. I feel all undressed, and at the same time sort of relieved! I can't get my feelings straight. How did I get into this mess?

JG: We're finding that out: How you learned what you learned, how you rehearsed your way of being the way you are, how it's worked for you until now, and what it costs you to keep it up.

Janice: That's a lot to sort out.

RLP: You can do it. The key is in seeing that it's a way. It's not a necessity. It's a way that you worked out. Apparently the purpose at the center of it was to find a position among others in which you could be innocent, and they would be responsible. To grow up is to learn that there ain't no such place!

Janice: I think I'm getting it.

RLP to JG: O.K. Then let's read these memories as if they were little fables or parables, little stories each of which has a moral, and see if we can get the pattern of their morals into clear statements. Then, if our statements make any sense, we'll see whether it's useful to Janice [turning to Janice], whether you can use it to begin reconsidering things. O.K.?

Janice: O.K.

Early Recollections Summaries. Promptly upon completing the review and interpretations of Janice's early recollections, it was time for the therapists to consolidate the meanings uncovered in the discussion of the data, and to put these ideas forward in statements in such a way as to be comprehensible and useful to the client.

The consulting therapist then dictated two summaries of the subjective sense expressed in the early recollections: *The Pattern of Basic Convictions* and *The Mistaken Ideas*. Janice listened as the recording therapist wrote down the dictation. These papers were later printed and given to the client.

Turn to Chapter 14 to review in more detail our analysis of the discussion of Janice's early recollections. This analysis illustrates, by way of example, how the content of the two final summaries of the lifestyle assessment may be arrived at and composed. The two summaries in the Case of Janice appear on p. 396 and p. 405.

Other Applications of Early Recollections

In addition to their place in lifestyle assessment, early recollections may be used at any time in the course of therapy, particularly at moments of agitation, by asking the client, "When was the first time in your life you felt this way?" The early recollection that arises will be of an incident that, while it may have taken place long ago in childhood, has a meaning governing client movement right now, in the present day. The early recollection will reveal an underlying attitude or assumption about life, others, and the world that may be distorted or erroneous. Calling up a recollection in a moment of disturbance offers client and therapist the opportunity to explore a *basic conviction* that may be preventing the client from moving forward comfortably in life and with others.

Early recollections may also be part of an *interim* or *termination interview* to measure the extent of a client's changes in personal growth, shifts in outlook, and development of maturity. To accomplish this, we inquire again into the client's store of remembered incidents from childhood up to age nine or ten. We then record *two sets* of ERs, *unprompted* and *prompted* ERs, as follows.

First, we record a collection of *unprompted* ERs gathered without reference to the original ERs, recorded in the same manner as were the originals in the lifestyle assessment. (If in this process, the client brings up an ER already recounted either in the family material or in the original set of ERs, ask the client to tell the story again.)

We then record a second set of *prompted* ERs. With the original set of ERs in hand, we remind the client by a word or phrase of the stories in the original set, and make a record of the current version. (If, as sometimes happens, the client has already recounted one or more of the original set among the *unprompted* ERs, move on to the next original ER with a prompting word or phrase.)

ERs are understood as expressions of the personal frame of reference, and personal orientation in life. In the economy of a very few recollections reviewed and interpreted in the course of therapy as an interim assessment or at therapy's termination, a reliable sense may be gained of how and to what extent changes have taken place in the pattern of an individual's unique style of living. (For a format for collecting the prompted and unprompted recollections and for summarizing the *Shifts in the Pattern of Basic Convictions*, see the *IPCW*, pp. 33-39)

Conclusion

With the conclusion of Chapter 11, we also reach the end of Part III, in which we have addressed the process of reviewing and interpreting the client's contemporary life situation, the family material, and the client's early recollections.

The final task in both the initial interview and lifestyle assessment procedures is the presentation of summaries of the material to the client. Part IV, "Summaries of Understanding," comprising Chapters 12, 13, and 14, focuses on methods for the development and composition of these summary statements.

Exercise for Chapter 11.
Early Recollections
(Use in dyads, small groups, or class, or with clients.)

1. Write down your earliest recollection (not a report), focusing on a single incident, up to age nine or ten. Consider where you were, who was there (if anyone), and what happened. Narrow the memory to the most vivid moment: How did you feel in that moment? Recount two more memories in the same manner.

2. Interpret your ERs taking into account issues outlined under "Key Categories to Consider in Reviewing and Interpreting ERs" in the present chapter. As you review and interpret the ERs, reflect on what they express *relevant to your present situation:* What do they reveal about your current convictions, biases, attitudes, opinions, and evaluations regarding self, others, and the world?

PART IV: SUMMARIES OF UNDERSTANDING

The [lifestyle] in its entirety can be kept intact only if the patient succeeds in withdrawing it from his own criticism and understanding. . . . [We] must grasp the special structure and development of that individual life with such accuracy, and express it with such lucidity, that the patient knows he is plainly understood and recognizes his own mistake.

Alfred Adler (pp. 334-335)

INTRODUCTION TO PART IV

This final section of our text is devoted to the initial interview and lifestyle assessment summaries. The summaries are composed in the client's presence by the consulting therapist at the conclusion of each review session, as oral statements that the recording therapist takes down in a verbatim written record for later transcription. Such oral summaries are also possible for therapists working alone, who can record the statements.

It is usually deeply affecting and valuable to the client to hear the summary of all that has gone before spoken aloud, on the spot. In the words of one client, "When I was listening, I was mesmerized. Then I felt profound relief. I don't know . . . I just thought, 'My story makes some kind of sense . . . not only to me, but also to them, to others.'"

The direct presentation of the summaries to the client follows the model taught by Dreikurs, who pioneered the method, and who insisted that anyone could learn how to do it.

Granted Dreikurs's certainty that anyone can learn how to do it, the procedure continues to give pause to many otherwise self-confident clinicians. Their hesitation seems to be based on an impression either that summaries are intuitive, even oracular utterances, issuing out of a depth of inspiration inaccessible to the uninitiated; or, that they are spontaneous, following no form, and so requiring no preparation.

Part IV is designed to disabuse readers of such impressions by demonstrating that, on the contrary, summaries are disciplined, studied compositions resting upon a structure and employing a method that can be mastered, a method whose content and cogency relies not so much upon oracular wisdom or quick wits as upon attention.

The method presented here for composing summaries depends for its success upon its use by the clinician as a frame of reference borne in mind throughout each review session for assessing information, and for considering how each item fits into the pattern of a coherent, comprehensible whole. Adler (1964) is clear about the challenge:

> The foremost task of Individual Psychology is to prove this unity in each individual — in his thinking, feeling, and acting, in his so-called conscious and unconscious, in every expression of his personality. . . . This unity we call the style of life of the individual. (p. 175)

We believe that the ability to locate this unity, and to summarize the themes of a self-consistent lifestyle in a way that is useful and encouraging to the client, is within the scope of every empathic, well-educated, and professional counselor's competence.

The methods we use in our practice and recommend in this text have been refined over many years with many different people. That does not mean that we claim our way to be the only or the right way to "prove [the] unity [of] each individual." Once practiced in the art and science of the understanding of persons (G., *Menschenkenntnis*) each therapist will develop a technique, a format, and a vocabulary consistent with the unity of his or her own style.

Chapter 12. IMPRESSIONS DERIVED FROM THE INITIAL INTERVIEW

The material in this chapter refers to the data gathered in the initial interview inquiry (Items 1-36, pp. 43-92), and whatever additional information emerges in the discussion between the client and the two therapists during the review and interpretation of the data.

The text of this chapter is divided into several sections. First is a list of nine topics to be considered when composing the initial interview summary. Second is "The Case of Joan, with a précis of the interview material and with a complete account of the *Summary of Impressions Derived from the Initial Interview*, which is presented and analyzed to illustrate the application of the nine topics in a single case. "The Case of Peter" is then similarly presented and analyzed. After Peter's case come two additional cases, presented more briefly, with each interview in précis, followed by the *Summary of Impressions* in each case.

The topics listed below for possible inclusion in a summary appear in a suggested order, but this sequence is not to be regarded as ordained. Priority and sequence are questions of emphasis and style arising from each therapist's clinical judgment and sense of language. Readers should also bear in mind that not every summary will address all topics, some of which will be relevant and useful in one case and not in another.

Topics to Consider for a Summary of the Initial Interview

1. The presenting problem

2. Onset and history

3. Exogenous factor

4. Childhood rehearsal of place, especially as it may be limiting the client's success in dealing with the current problem

5. Masculine and feminine guiding lines bearing upon the client's faulty perception of self and others (convictions, expectations, ambitions, consequences)

6. Relevance of the Big Numbers to the client's complaint

7. The client's resources for meeting the difficulty; examples

8. Reinterpretation of the complaint (the client's role; the roles of others)

9. The task to be addressed

The Case of Patti

Précis of Data from the Initial Interview

Patti is a competent forty-three-year-old sales representative for a company supplying linens to hotels and institutions. She came to the initial interview with two issues: First, she was concerned about her deteriorating relationship with her boss; she said he was "picking on" her and criticizing her, in spite of her successes with the company over many years, and that he seemed to have made an alliance recently with another woman staff member, displacing Patti as his favorite. Second, three months before our meeting she had learned that her husband of twenty-three years had "fondled" their oldest daughter, Jennie, age seventeen. Patti learned about her husband's misbehavior from Jennie's boyfriend when the two young people attempted to elope. The boyfriend stated that he had tried to marry Jennie without the family's knowledge only because he wanted to get Jennie away from her father. A confrontation with the husband ensued. After first denying the activity, and then attempting to blame it on his daughter ("she enticed me"), the husband admitted it and has subsequently been in counseling. "My daughter has forgiven him, but I can't," Patti says. "I cry all the time, and can't feel close to him any more." She feels, further, that she let her daughter down by not realizing what was going on. She feels betrayed both by her husband and by her boss.

The Case of Patti (continued)

Summary of Impressions Derived from the Initial Interview

Patti presents herself as a woman who has been waking up out of a long sleep over the last three months, aroused by a sense of having been betrayed by the men in her life. Until now she has seen herself as a weak woman in a world of strong men. Her present confusion arises because she has now begun to see herself as a strong woman in a world in which there are many weak men.

Because until now — owing to her childhood training with a powerful father and an ineffectual mother, she has made the mistake of thinking that her strength could only be real if it were sustained by, approved of, and recognized and rewarded by a man — she has been kept from a clear awareness of her strength.

Her father, her husband, and her boss have all operated as if they were intimidated by the very idea of the possibility of a competent woman's domination (perhaps because each of them leaned on women and counted on them for comfort and support).

Her disillusionment over her husband's misconduct with their daughter, Jennie, has helped to open her eyes, and the current situation with her boss is helping to open them wider.

She has an enormous advantage in that she is only forty-three years of age, and no matter what happens now, she has already seen worse; and in that, by the time she is forty-five (the age Mother was when Father left Mother), Patti will have had the opportunity to decide what she wants in life, and will have acted to secure it.

Furthermore, she has already made a choice to discover how much she can do as one of the New Women in a New World, and she can take advantage of the other great advantage she enjoys as a result of her long training in cultivating an interest in the interests of others.

With her eyes open, she can see that her elevation of men into the position of superiority is a diminishment of herself as a woman; and further, that it places an intolerable burden on the men in her life, who must struggle to live up to her exaggerated expectations of them, and so are precluded from having an equal, caring relationship with her.

Her task now is to get a clear understanding that ideas of "strength" and "weakness" are fictions, and to see more clearly that her effectiveness is a function of choice (not of strength or gender), of choosing and refusing what she will do and what she won't do in any situation, including what she will think about it and how she will feel about it.

The Case of Patti (continued)

Method for Composing the Summary of the Initial Interview

Topic 1. *The presenting problem*

State the problem clearly, using the client's own language wherever it may be appropriate. Patti's presenting problem is stated in paragraph one:

> Patti presents herself as a woman . . . aroused by a sense of having been betrayed by the men in her life. . . . Her present confusion arises because she has now begun to see herself as a strong woman in a world in which there are many weak men.

Topic 2. *Onset and history*

It is helpful to provide a sense of the temporal context in which the client has become aware of the difficulty. Some features of this context may be lifelong; others may reach back only through a brief period to a recent moment of crisis.

Sometimes there is a combination, as represented in the following opening sentence of a summary done for Paula, a woman of fifty-two: "Paula presents herself at the end of a period of discouragement going back at least three years, in some ways nine years, and probably in some other ways, fifty-two years." This was a person who had experienced severe setbacks three years before and nine years before, and who had said, during the inquiry, "I've always felt out of place."

About Patti, we state in paragraph one:

> Patti has been waking up out of a long sleep over the last three months.

Topic 3. *Exogenous factor(s)*

These factors arise outside the scope of the person's rehearsed style, and therefore may present a challenge for which the person is unprepared. The resulting crisis is a faltering, a breakdown of the style, which is now revealed as ineffective. (If this revelation is sudden enough, and far reaching, a radical loss of orientation can result, and the person may be described as suffering a "nervous breakdown," or even a "psychotic break.") Here, in paragraph four, we tried to put a positive meaning on Patti's disorientation:

> Her disillusionment over her husband's misconduct with their daughter, Jennie, has helped to open her eyes, and the current situation with her boss is helping to open them wider.

Topic 4. *Childhood rehearsal of place, especially as it may be limiting the client's success in dealing with the current problem*

Topic 5. *Masculine and feminine guiding lines bearing upon the client's faulty perception of self and others (convictions, expectations, ambitions, consequences)*

References to these two topics are *intertwined* in Patti's summary because her training and self-training were shaped around her deference to the perceived power of the man (her masculine guiding line), and her dismissal of Mother as "ineffectual" (her feminine guiding line).

Her convictions regarding gender therefore appear to be that men are indisputably powerful while women are inherently weak. Her consequent expectation, governing her own movement in life as well as her sense of what she can expect from others of both sexes, is that women are coming from behind and are precluded from catching up to the men, no matter what their talent, brains, or effort. So, also in paragraph one:

> Until now Patti has seen herself as a weak woman
> in a world of strong men.

The ambition that flows out of these convictions and expectations was formed at an early age: To be effective, she must be under the protection of a man. So, paragraph two:

> Because until now — owing to her childhood
> training with a powerful father and an ineffectual
> mother, she has made the mistake of thinking that
> her strength could only be real if it were sustained
> by, approved of, and recognized and rewarded by a
> man — she has been kept from a clear awareness of
> her strength.

In paragraph seven of the summary, we draw Patti's attention to the consequences of these convictions, expectations, and ambitions, as to something she is now free to ponder and understand, and in doing so we begin to make it more difficult for her to proceed in innocence, should she persist in her erroneous way of thinking:

> With her eyes open, she can see that her elevation of men into the position of superiority is a diminishment of herself as a woman; and further, that it places an intolerable burden on the men in her life, who must struggle to live up to her exaggerated expectations of them, and so are precluded from having an equal, caring relationship with her.

Topic 6. *Relevance of the Big Numbers to the client's complaint*

During the review and interpretations, the therapists and the client note and consider any relation between the current experience of difficulty and the client's privately held expectations regarding age. We have referred to these temporal features of the individual's phenomenological field as the Big Numbers (see Chapter 6). These numbers relate to events or turning points in the histories of the parents. They are attended to as features of the gender guiding lines indirectly, and at low levels of awareness, sometimes even unconsciously. It is important to recognize that meanings given to these numbers are always *fictional*, even when they may coincide with numbers or ages whose meanings are generally agreed upon in the culture (such as, "sweet sixteen"). In Patti's case there was a clear connection between her expectations regarding her age and her feminine guiding line. Here is how we noted it, in paragraph five of the summary:

> She has an enormous advantage in that she is only forty-three years of age, and no matter what happens now, she has already seen worse; and in that, by the time she is forty-five (the age Mother

was when Father left Mother), Patti will have had the opportunity to decide what she wants in life, and will have acted to secure it.

Encouragement is the *sine qua non* of all effective psychotherapy or counseling. It has at least as important a place in the work of the initial assessment as it does in any subsequent transactions. We therefore seize every opportunity the client gives us to introduce it at the beginning of our work together. For example, by stating, as we did in the above phrase, that Patti will have her life sorted out as she wants it, and that she will do so by experiencing herself as a chooser and an actor within a short period of time. We have lent support, by rewording it, to an expectation she had confided in us. Then, by recording it as a prediction, we added to its force, and to her confidence in it.

Topic 7. *The client's resources for meeting the difficulty; examples*

Part of the responsibility of the therapist is to aid the client in recognizing and developing useful resources. In a time of crisis the lifestyle, which until now has enabled an acceptable degree of successful adaptation, is revealed as being inadequate to the demands of the situation. Clients complain, "I don't know what to do! I can't cope!" Even so, everyone has some resources of courage, intellect, or other capacities which may be hidden from awareness or not clearly understood and valued, especially in a time of disorientation. Two of Patti's resources, which can serve her in her present difficulty, are brought directly to her attention in paragraph six of the summary:

> She has already made a choice to discover how much she can do as one of the New Women in a New World, and she can take advantage of the other great advantage she enjoys as a result of her long training in cultivating an interest in the interests of others.

Patti was encouraged by the thought that entering a career and making a success of it were choices she had made, and that the courage with which she had met the challenge of occupation was available to her for meeting the current challenge. She had not appreciated the significance of this accomplishment, not having seen it as an index of courage and determination. Further, she had not experienced her success as a salesperson as demonstrating her "interest in the interests of others," though when it was mentioned she immediately recognized what we meant. She had not valued this in herself, and so had not realized how it might assist her now in offering empathy and friendship to her ashamed and unhappy husband.

Topic 8. Reinterpretation of the complaint (the client's role; the roles of others)

It was important for Patti to see her role in the difficulty, as set forth in paragraph seven of the summary (her elevation of men, and the consequent burden she placed on them). It was also important for her to consider how the men in her life might be perceiving her (and other women) and how an awareness of those perceptions could help her to understand their movement, especially in displays of power and dominance (Father toward Mother; husband toward daughter). "Could it be," we asked, "that these men, who clearly depend on the women to take care of them and their interests, are fearful that unless they act this way they might be subject, at best, to the loss of the women's respect and service; and, at worst, to the power and domination of the more capable women?" This idea is recorded in paragraph three of the summary:

> Her father, her husband, and her boss have all operated as if they were intimidated by the very idea of the possibility of a competent woman's domination (perhaps because each of them leaned on women and counted on them for comfort and support).

Topic 9. *The task to be addressed*

Typically the summary closes with a statement of the task needing to be addressed by the client. In paragraph eight of the summary we set forth the two major ideas which she will have to reconsider: her fictions of strength and weakness, and her faulty convictions of being unable to act effectively on her own:

> Her task now is to get a clear understanding that ideas of "strength" and "weakness" are fictions, and to see more clearly that her effectiveness is a function of choice (not of strength or gender), of choosing and refusing what she will do and what she won't do in any situation, including what she will think about it and how she will feel about it.

The Case of Peter

Précis of Data from the Initial interview

Peter is a brilliant thirty-six-year-old public relations consultant. He came to the office in the aftermath of a divorce, after twelve years of marriage. It had been the first marriage both for him and his wife; they had adopted two children, girls now six and eight years old, after they learned that his wife was unable to conceive. Peter's mother is living and in good health. Father died at the age of thirty-seven when Peter was ten years old. When Peter was eleven years old his older sister had a nervous breakdown, and has been in institutions off and on ever since. Before these difficulties he had been the baby of the family. As he put it, "At ten I became the man of the family, and, at eleven I had to fill the place of both of the children as far as my mother was concerned." Peter's wife initiated the divorce after disclosing to him that she had been having an affair with his best friend for a year and now wanted to marry the other man. At the time of our first interview, the former wife and former best friend had been married for six months. Peter had been depressed during all of that period, and was seeing a psychiatrist who medicated him. "It was some help, but it didn't get to the trouble, and I don't feel any better." Prior to the start of his wife's affair, Peter had suffered a business failure when another "friend" absconded with the funds of their partnership, leaving Peter with numerous financial obligations which required all of his assets to repay. He has regained his financial position and is now once more a wealthy man. His present employer says that his job is secure "no matter what you have to do to get well." His circle of friendships is limited to married couples, friends from his marriage who "have stuck by me." He is devoted to his children and lets nothing interfere with his visitation rights. After eighteen months of job demands, child care, and commuting to and from his apartment, his place of work, and his children's primary residence, he is overextended, lonely, and exhausted.

The Case of Peter (continued)

Summary of Impressions Derived from the Initial Interview

Peter presents himself at a time in his life when the question of his vulnerability to being used and abused is uppermost in his mind. He is dimly aware of an urgency to settle this issue while he still has time, because his father died of heart trouble at the age of thirty-seven, just one year older than Peter's present age, and Peter's heart is troubled.

He knows himself to be devoted to doing whatever is necessary for anyone whom he allows to count upon him, as he did for Mother after Father's death, as he has done for his sister since she was fifteen-years-old (when she returned to the family after her first stay in the hospital), and as he did for his creditors at the time of his business partner's betrayal. In all these works he is uncomplaining, and it is by these works that he recognizes himself as a Good Man.

He is not clear about his hidden expectation that his goodness will be rewarded by the gratitude and adoration of others, and that he will earn this reward for his fidelity in the face of difficulty. He maintains this expectation in a kind of secret, one-sided contract with life, as if he hoped that by keeping it secret the other side would feel uncoerced in entering into it, seeing his virtue as irresistible, and wanting to do its part in the bargain without knowing that there is a bargain.

His version of the contract goes something like this: If I am good I will get what I deserve. If I demand what I deserve, I will no longer be good. Therefore, I must not demand what I deserve, because then I would no longer deserve it. When I do not get what I deserve, the only thing left open to me is to look around for ways to be even more good, to redouble my efforts, and gently to reproach the others for their failure to keep their part in the bargain by confronting them with my undeserved sadness.

The brilliance and ingenuity so prized by his professional colleagues (brought to bear earlier to solve problems surrounding the collapse of his private business venture) will assist him in solving the riddle of his goodness. He will begin to see that goodness is only good when it can be measured in terms of usefulness, and that it is no longer useful for him to suffer.

He is beginning to realize that he has struck a bad bargain, and may decide to end an interpretation and a way of life that has been destructive for him. He may decide to do that now, beginning to live a life of honest self-assertion at about the same age his father was when Father's life ended.

The Case of Peter (continued)

Method for Composing the Summary of the Initial Interview

Topic 1. *The presenting problem*

Peter came to see us because of a depression that had been medicated but not remedied. In the course of the discussion we discovered with Peter that he felt overworked in the service of others and then mistreated by them. We therefore defined the presenting problem as vulnerability (expressed in depression) in paragraph one of the summary:

> Peter presents himself at a time in his life when the question of his vulnerability to being used and abused is uppermost in his mind.

Topic 2. *Onset and history*

Peter's story revealed a pattern characterizing his style from age ten, the time when his father died, and coalescing at age eleven when his sister defaulted, only one year after the death of her favored parent. In this period he moved from the position of baby to that of Man, and from one of the children requiring Mother's attention to the central role in Mother's life. From this point on he took up his role as the one who could be counted on. His motto could be, "I stay the course." Paragraph two of the summary states:

> He knows himself to be devoted to doing whatever is necessary for anyone whom he allows to count upon him, as he did for Mother after Father's death, as he has done for his sister since she was fifteen-years-old (when she returned to the family after her first stay in the hospital), and as he did for his creditors at the time of his business partner's betrayal.

Topic 3. *Exogenous factor(s)*

The current crisis in Peter's life has been facing him for about a year. It was occasioned by the end of his marriage, following upon the simultaneous betrayals he suffered at the hands of his wife and his best friend who, six months prior to our first meeting, had married each other. By the time of this initial interview it had become clear to Peter that his former wife's new marriage was stable, and that she would not be coming back to him. All hope of his reconstituting his family on its prior foundations had ended. Paragraph one of the summary encompasses this awareness in its last phrase ". . . and Peter's heart is troubled."

Topic 4. *Childhood rehearsal of place, especially as it may be limiting the client's success in dealing with the current problem*

Peter made a place as the Good Baby, Good Child, and Good Man. There are many Good Boys who become Good Men, but he would be the Best of the Good. As in the last line of paragraph two of the summary:

> In all these works [caring for Mother, Sister, creditors] he is uncomplaining, and it is by these works that he recognizes himself as a Good Man.

In order to maintain the place of the Good Man, Peter cannot consider his part in bringing about the difficulties he faces in his life. Were he to see the overreaching and condescension in his style, he would have to see his un-goodness, losing his Good Man's innocence. And so the one-sided deal he has made with life must remain hidden.

Topic 5. *Masculine and feminine guiding lines bearing upon the client's faulty perceptions of self and others (convictions, expectations, ambitions, consequences)*

As we discussed with Peter (and as alluded to in Topic 4), the task he has set for himself in life is in line with a picture of a man who overcomes difficulties at any cost; the task he has assigned to the others (especially women) is to be grateful and to admire him for these efforts. These ideas are expressed in paragraph three of the summary:

> He is not clear about his hidden expectation that his goodness will be rewarded by the gratitude and

> adoration of others, and that he will earn this
> reward for his fidelity in the face of difficulty. He
> maintains this expectation in a kind of secret,
> one-sided contract with life, as if he hoped that by
> keeping it secret the other side would feel
> uncoerced in entering into it, seeing his virtue as
> irresistible, and wanting to do its part in the bargain
> without knowing that there is a bargain.

Because roles defined in this way ("I will be good and never falter; you will admire me, and give me whatever I want") are untenable in commonsense social transactions, Peter's privately defined contract has remained un-understood by him; and, since goodness is by definition unassailable, his mischief has remained un-understood by others, even when they have chafed under the arrangement. Until now, the bargain has been inaccessible to critical evaluation, standing unseen in the way of Peter's ability to reorient himself. He could, therefore, only continue doing what he knew how to do.

Topic 6. *Relevance of the Big Numbers to the client's complaint*

An awareness of age as it may be believed to reflect recapitulation along the gender guiding lines is worth exploring in any time of crisis as a source of reorientation for a disoriented person. Paragraph one of the summary states these connections:

> He is dimly aware of an urgency to settle this issue
> [his vulnerability to abuse by others] while he still
> has time, because his father died of heart trouble at
> the age of thirty-seven, just one year older than
> Peter's present age, and Peter's heart is troubled.

Topic 7. *The client's resources for meeting the difficulty; examples*

Paragraph five of the summary calls upon qualities that are available to Peter in sorting out his (one-sided) bargain with life:

> The brilliance and ingenuity so prized by his professional colleagues (brought to bear earlier to solve problems surrounding the collapse of his private business venture) will assist him in solving the riddle of his goodness.

Topic 8. *Reinterpretation of the complaint (the client's role; the roles of others)*

The problem he presented was depression; he came to see that the depression was a symptom of what we had discovered together, namely, his tacit (and one-sided, secret) agreement with life, as we state it in paragraph four:

> His version of the contract goes something like this: If I am good I will get what I deserve. If I demand what I deserve, I will no longer be good. Therefore, I must not demand what I deserve, because then I would no longer deserve it. When I do not get what I deserve, the only thing left open to me is to look around for ways to be even more good, to redouble my efforts, and gently to reproach the others for their failure to keep their part in the bargain by confronting them with my undeserved sadness.

During the review and interpretations of the initial interview data, Peter and the two therapists discussed the idea that his depressive symptoms seemed to arise, ultimately, from an old conviction that he had been a boy sent (unfairly) to do a man's work. It was (we considered) also possible for him to believe that life had presented him with opportunities to take his place as a man (escaping the position of baby) and that he had readily grasped this challenge.

In paragraph six of the summary, we acknowledge his willingness to change the bargain he has made with life:

He is beginning to realize that he has struck a bad
bargain, and may decide to end an interpretation
and a way of life that has been destructive for him.

Topic 9. *The task to be addressed*

In paragraphs five and six of the summary, we identify his tasks,
noted above, and express our confidence that:

He will begin to see that goodness is only good
when it can be measured in terms of usefulness, and
that it is no longer useful for him to suffer.

The final paragraph of the summary refers once again to his
father, in order to encourage him to think of the timeliness of his
present opportunity:

He . . . may decide to [end the bargain] now,
beginning to live a life of honest self-assertion at
about the same age his father was when Father's life
ended.

Having completed "The Case of Patti" and "The Case of Peter,"
we close this chapter with a précis of data and the *Summary of
Impressions Derived from the Initial Interview* in two additional
cases: "The Case of Andrea" and "The Case of Gary."

The Case of Andrea

Précis of Data from the Initial Interview

Andrea is twenty-three years old, currently between jobs. Since
graduating from college, she has worked for an investment firm
and for an arts organization. In neither of these jobs did she "catch
fire," though her bosses and colleagues let her know that they liked
her work and wanted her to stay. Father is a successful
entrepreneur, and Mother is a successful sculptor and teacher of
sculpture. Her only sibling, a brother two years older, is an

architect who has just been asked to join a firm with an international reputation. She was Father's favorite, and was brought up by him to "see the world through rose-colored glasses. He said I could be anything I wanted to be." Andrea has been depressed for two months, since going to her home town for the Christmas holidays. "I saw all my friends, and they wanted to know what I was doing with my life." About her goals, she said, "I want to get out there and work, but part of me says I want it handed to me on a silver platter." She feels that her "circle" wonders why she isn't married, doesn't even have a suitable man in her life, doesn't have a prestigious job, and doesn't even seem to have an "important" group of friends and activities in her new setting. She says that her friends seem to resent and envy her money, looks, and talents. "With all I've got going for me, I should be able to do anything. I should be a saint! But I feel so undeserving. I should do more. I shouldn't have any problems. I've become a recluse. All I do is go to the movies and eat." She says that on the first impression people enjoy her and find her personable. Over time, "They find I'm the least confident of people." Father keeps telling her, "It's going to be fine." She has doubts. "I try really hard, and I can't even remember what it feels like to feel successful."

The Case of Andrea (continued)

Summary of the Impressions Derived from the Initial Interview

Andrea presents herself as a young woman burdened with a combination of grand ambitions mixed with a more limited supply of self-confidence.

As the second child of ambitious parents, and the only girl, she appears to have grown up with a sense of being special, and with a feeling of being less than whatever it is she thinks is expected of her.

Her enjoyment of her advantages is undermined by her exaggerated standards of fairness ("with all I've got going for me, I should be able to do anything"), and her sense of having to do something spectacular in order to justify the encouragement given to her and the support she receives.

Her creativity adds to her burden, since it increases the likelihood of her being seen as special, and therefore increases her sense of obligation to achieve in special ways, to live up, to catch up, and to be up.

The depressive reaction since last December appears to have come at a time of restlessness, when she sensed herself to be out of the line of movement, and in some ways even backtracking to a position of financial dependence on her parents.

Her training as the secondborn in this family of achievers may have predisposed her to restlessness, overambition, and a feeling of coming from behind in a field of thoroughbreds. This would make her current situation all the more onerous for her, leaving her vulnerable to a kind of pessimism about being able to get on with things soon enough and well enough to "deserve" the confidence of her family and her friends.

All these difficulties can be seen as advantages. Her task is to redefine her exaggerated ambitions, and to understand her "guilt" over being undeserving as an unwillingness to rest with what she has long enough to discover what would happen if she were to get on with what she can do with it.

The Case of Gary

Précis of Data from the Initial Interview

Gary is a man of twenty-eight who has been disoriented in life since an automobile accident two years before disabled him in such a way as to preclude his continuing in his chosen profession as a musician. (He had fallen asleep at the wheel during a long, soul-searching drive he took to ponder whether to defy his father's advice to keep music an avocation while pursuing a career in business more likely to secure an ample income.) He is on a leave of absence from his present employer, a real estate development firm, because of continuing physical problems, this time a severe rash. Gary's physicians told him that the rash has no discernible physical basis and suggested that he see a psychologist. Because the accident was disfiguring, Gary also brings up his sense that "I can't give myself to anyone any more. I can't reach out. People won't accept me." He contrasts his present situation with his boyhood: "I was a sunny child; always had a smile on my face. I didn't talk about my problems; just the good stuff." He says, "I don't shove my problems off on others" as an adult. Recently his girlfriend of several years ended the relationship. "I honestly think it was because I couldn't talk about my problems. I kept it all inside. I just couldn't marry her. It wasn't fair to her. I don't know what is going on in my life. How could I ask her to get into such a mess?"

The Case of Gary (continued)

Summary of Impressions Derived from the Initial Interview

Gary presents himself at the end of a long period of disorientation which began at the time of his accident, that is, not *since* the time of his accident, but *at* the time. His sense of himself had, until then, been associated with his musical ability and his successful career in music; and his desire to follow the line of this career in the life of a professional musician seemed to require of him that he abandon his role as the "sunny child" in his family, and to assert his desires in opposition to his father's "good advice"; it therefore appears that he began, at that time, a struggle to reorient himself in a new role.

His consequent sense of disorientation contributed to the recent ending of a long love affair, in that he hesitated to include another person in his uncertainties. His unhappiness in the world of real estate adds to his lack of clarity about the next step in his life.

Until now he has been able to preoccupy himself with his skin disorder, and his search for a medical solution. That he is now willing to reconsider the purpose and function this disorder may serve in his life is already a sign of optimism, and of his courage.

As the baby of his family, Gary enjoyed a place of prominence. All that was required of him was to be a source of joy. The bright side of this assignment was the pleasure he could take in bringing pleasure to others; the dark side was that he felt disallowed from having any desires and preferences of his own, much less any problems or complaints. He had no training, therefore, for the forthright confronting and solving of problems with the cooperation and interest of others. He grew up with an uneasy feeling that if he admitted to any difficulties, his special place would be lost to him.

The maintenance of the rash may make it possible for Gary to delay addressing his current distress over the need to set a career path for himself. It may be that he thinks that "until this problem is solved, I can't get on to anything else," such as the pain of his loneliness or the uncongeniality of his occupational opportunities.

At the age of twenty-eight, Gary is entitled to reconsider his training for solitary problem solving, and to see it as an exaggeration of normal self-reliance and responsibility. Since he took this on at an early age, he will have to learn all over again how to accept the invitation extended by his family and friends, and how to reorient himself around the idea of accepting their love, care, and willingness to struggle by his side. It is Gary's self-imposed solitary confinement that is disfiguring, not his physical "confinement," and he can release himself from it in an acceptance of his solidarity with others.

Conclusion

In Chapter 12 we have discussed topics to consider in the composition of an initial interview summary, and have clarified the intent of each of the suggested topics, with illustrations from cases.

In our practice the initial interview often serves as a form of brief psychotherapy. A number of clients are satisfied with what we uncover together here, and decide that further therapy would not be appropriate or useful to them at this time in their lives. We recommend at least a single session to follow up on the things discussed in the initial interview. Most agree, and come in again the following week. Others do not, though we may hear from some of them again after as much as a year or more has passed. Sometimes it is only after a series of crises (such as the one that led to the initial interview) that a person is prepared to enter into the psychoclarity process on a continuing and disciplined basis.

"An individual's lifestyle is developed as his own unique manner of coping with life's problems, according to the way he feels and sees them in his particular picture of the world," Adler wrote (1979, p. 111). When the client's lifestyle no longer serves to cope, that is, when the person's unique way of feeling and seeing is no longer effective in addressing life's problems of love, work, and community, completing a lifestyle assessment may seem more clearly advisable to the client, and we can return to it then.

In the next chapter we turn to that part of lifestyle assessment in which we develop the *Summary of the Family Constellation.*

Exercises for Chapter 12.
Impressions Derived from the Initial Interview
(Use in dyads, small groups, or classes, or with clients.)

In a current or past case, or a case you are studying:

1. Define the *presenting problem*, preferably in a single declarative sentence.

2. Identify the *exogenous factor* in the case.

3. Provide one or more examples of the client's *proven resources* for meeting the current challenge.

Chapter 13. SUMMARY OF THE FAMILY CONSTELLATION

Adler (1970, p. 80) used the term "family constellation" to designate the system of relationships in which self-awareness develops. This system includes and is maintained by oneself, one's parents and siblings, and any others living in the household of the family of origin (or otherwise recognized as part of its extension). The term represents the family by analogy to a constellation in astronomy: a group of bodies, each of which moves and has a place in relation to the movement and places of the others.

Following Adler, Dreikurs (1973) wrote:

> A clear formulation of a person's lifestyle can be obtained through investigation of his family constellation, which is a sociogram of the group at home during his formative years.

> This investigation reveals [the individual's] field of early experiences, the circumstances under which he developed his personal perspectives and biases, his concepts and convictions about himself and others, his fundamental attitudes, and his own approaches to life, which are the basis for his character, his personality. (p. 87)

The family provides the context for the primary rehearsal of a way of movement among others which is to become the individual's style of living. Just as the individual is not encapsulated in "self-boundedness" (G., *Ichgebundenheit*) (Griffith & Powers, 2007, p. 90) and cannot be understood in isolation from this context, the family cannot be encapsulated as a system apart from a community of others in a given time and place in history. Therefore, we also consider and include in the family constellation summary those impressions and details of linkages to

351

the world outside the immediate family that were factors in the client's training and self-training. These include the extent of involvement in and interaction with members of the extended family; role models outside the family; experiences in the social life of the neighborhood and the school; and impressions and conclusions relating to the ethnic/racial, religious, social, and economic milieu reported by the client in the inquiry and in the review and interpretations.

Remember that a summary is a summary; that is, according to the *American Heritage Dictionary* (1992), "a presentation of the substance of a body of material in a condensed form" (p. 1798). The family constellation summary is a brief statement, incorporating major points. The summary is not intended to be a biographical account of the client's experience as a child and adolescent; rather, its purpose is to give a fresh expression to the client's *subjective sense* of those years as they relate to the unique style of movement the client learned in the beginning, and continues to practice in the present.

In short, the family constellation summary facilitates the client's respectful self-understanding for the way the client created a style of living, offered in the summary in a way that is *encouraging* for the client in the current situation (as with all therapeutic work in Individual Psychology).

In this chapter we present a method for composing the *Summary of the Family Constellation* based on the twelve topics below, with suggested sentence stems for each topic on pp. 354-355. Each topics is then reviewed in detail and illustrated with case material excerpted from actual family constellation summaries. The chapter concludes with an analysis of the *Summary of the Family Constellation* in "The Case of Dan." (Dan's case is presented in its entirety in Chapter 4; for the summary, see pp. 117-120.)

Topics to include in the *Summary of the Family Constellation*:

1. Birthorder

2. Family atmosphere

3. Family values

4. Masculine guiding line

5. Feminine guiding line

6. The place made by the client in the family

7. The places made by each of the siblings

8. Other images: role models and alliances

8. The experience of neighborhood and school in childhood and puberty

9. The experience of puberty

11. The sexual challenge in adolescence

12. Major unresolved issues remaining from childhood and adolescence

Method for Composing the
Summary of the Family Constellation

The twelve topics are listed below, together with suggested phrases for use in composing the summary. Where information in the sentences is enclosed in brackets, follow the lead provided: Use your own client's information in that space. For example, the first sentence stem, relating to birthorder, is, "[The client] grew up as the [ordinal position], psychologically the [psychological vantage]." With client information written in, we have this sentence: "Sharon grew up as the third child in a family of four children, psychologically the baby of the first group of three with a much younger brother."

In the coming pages you will find many examples showing how to use each of the sentence stems we present below.

Topic 1. "[The client] grew up as [ordinal position], psychologically the [psychological vantage]. . . ."

Topic 2. "The family atmosphere was [description]. . . ."

Topic 3. "The family values emphasized by Mother and Father included [state and elaborate upon these common values]. . . ."

Topic 4. "The masculine guiding line set by Father stressed [state principal features of the masculine guiding line, especially those values not shared by Mother]. . . ."

Topic 5. "The feminine guiding line set by Mother stressed [state principal features of the feminine guiding line, especially those values not shared by Father]. . . ."

Topic 6. "[The client] made a place for himself/herself by [description of how client distinguished self in family: Consider activities, alliances, competitors, relationships with parents, and comparison of client's place to those established by siblings]. . . ."

Topic 7. "[Sibling's name], who was [state age or position relative to the client], made a place for himself/herself by [consider elements in Topic 6]. . . ."

Topic 8. "Other images presented to [the client] were [identify role models] who [description of their significance to the client]. [The client] made an alliance with [identify the person] who was important to [the client] because [describe relationship; include particularly any encouragement provided] and/or [the client] had an aversion to [identify the person] because [describe the relationship]. . . ."

Topic 9. "During childhood, [the client's] experience was [describe major features of self-assessment in neighborhood and school, including physical self-image]. . . ."

Topic 10. "Upon reaching puberty, [the client's] experience was [describe positive and negative features, including expectations about entering adult life, and what this would mean]. . . ."

Topic 11. "During adolescence, [the client] felt [describe the client's self-assessment] and met [note experience in meeting challenges of self-reliance, sexuality, and direction for life]. . . ."

Topic 12. "The question that confronted [the client] at the threshold of adult life remained: [Here pose a question that identifies the client's lack of preparation for adult success, including the conditions under which the client is prepared to go forward]. . . ."

Topic Review and Illustrations

Topic 1. BIRTHORDER

1. Name the client.

2. State the ordinal position.

3. State the client's psychological vantage vis-à-vis siblings.

Case Examples: Firstborn Only Children

- Laura grew up as the firstborn child of her parents, in fact, and psychologically, an only child.

- Matt grew up as the only surviving child after two failed pregnancies; psychologically an only child, and the only son.

- Don grew up as the firstborn son and only child of his parents, psychologically the second born of the two men in the family, with his father having the role of the firstborn son to a woman who operated as Mother to both "her men."

Case Examples: Psychological Only Children

- Scott grew up as the much younger of two and the only son, psychologically an only child with two mothers.

- Amelia grew up as the first of two children, psychologically an only child with a younger sister who, owing to her incapacity, was inconsequential to Amelia, except as a burden to be endured.

- Theresa grew up as the fourth of seven children, psychologically an only child who experienced herself as the "odd one out" between the older and the younger groups of children.

Case Examples: Firstborns With Siblings

- Pam grew up as the oldest of three girls, psychologically the firstborn daughter with two little sisters.

- Ernie grew up as the first of six children, psychologically the firstborn of the two older boys, all of whose four younger siblings remained in a group of their own.

- Doreen grew up as the first of two children, psychologically the firstborn daughter with a brother who was a firstborn son.

- Art grew up as the first of eight children, psychologically the firstborn son in a class by himself.

- Lisa grew up as the older of twin girls, psychologically the firstborn daughter with a younger sister.

Case Examples: Psychological Firstborn Children

- Rachel grew up as the younger of two, psychologically the firstborn daughter because of her older sister's disability.

- Clare grew up as the second child among five children, psychologically the firstborn daughter with an older brother and three younger siblings.

- Ben grew up as the second of three children, psychologically the firstborn son with an older sister who was a firstborn daughter, and with whom he shared a baby sister.

Case Examples: Secondborn Children

- Dick grew up as the second of five sons, psychologically the secondborn with three "kid" brothers and a firstborn "big" brother.

- Alice grew up as the second of six children, psychologically the secondborn of the first two children, who had entered the family before its "collapse," who enjoyed a "golden age" together, and whose shared memories kept them in a class by themselves.

- Jeffrey grew up as the second of two boys, psychologically the secondborn son, striving, not so much to catch up as not to fall farther behind.

- Bruce grew up as the second of three, psychologically the secondborn son with a younger sister.

- Patricia grew up as the second of two girls with three younger siblings. Her psychological vantage as a secondborn was severely dislocated upon the death of her older sister, which occurred when Patricia was five years old. She was disoriented both by the loss of her pacesetter, and by her felt demand that she assume the responsibility of leadership and example in the "first" position left open by her sister's death.

Case Examples: Middle Children

- Susie grew up as the second of four children, psychologically the middle child and only girl in a group formed by the first three, all of whom regarded the fourth child as a baby sister.

- Brian grew up as the second of three children, who eschewed the burdens of a firstborn son's position, and was psychologically the middle child and only boy.

- Sheila grew up as the first girl and second child in a family of five children, psychologically the middle child between an older brother and a younger sister.

- Andy grew up as the second son in a family of seven children, psychologically the middle child in the first group of five.

- Bob grew up as the third of five children, psychologically the middle child among four boys with a younger sister. This subjective sense of "middleness" was based in part on his awareness that the oldest of the boys was not making a place for himself successfully as the firstborn, and was being displaced by Bob's next older brother.

Case Examples: Thirdborn Children

- Sharon grew up as the third of four children, psychologically remaining the thirdborn with two older brothers, and refusing the position of the older of the two girls.

- Richard grew up as the third of three children, psychologically allied with his firstborn sister in their failed effort to keep his older brother, the secondborn child, from shining.

- Jamie grew up as the only girl with two older brothers, choosing to be a tomboy in alliance with her brothers, and remaining psychologically the third born of the children.

Case Examples: Youngest Children

- Abe grew up as the lastborn of six children, psychologically the youngest member of the group.

- Will grew up as the third of four children, psychologically the youngest of three boys with a younger sister.

- Anne grew up as the fourth of five girls, psychologically the youngest of the first group of four, all of whom shared a baby sister.

- Emily grew up as the third of three girls, psychologically the youngest, whose sisters took the positions of firstborn daughter and middle child.

Case Examples: Babies

- Warren grew up as the youngest of four children, psychologically the darling baby brother of them all.

- Louise grew up as the younger of two children and the only girl, psychologically the baby of the family.

- Norman grew up as the first of two children and the only boy, psychologically the baby of the family with a younger sister who took the position of firstborn daughter and "little mother" to "Mr. Norman," thus removing him from competition.

Topic 2. FAMILY ATMOSPHERE

1. Describe the family atmosphere set by the undertone of the relationship between the parents (or other caretakers resident in the childhood household).

2. Note any extraordinary changes in the atmosphere and the intervening life events relevant to such changes.

3. Note the ethnic/racial character of the family significant in establishing the atmosphere.

Case Examples: Family Atmosphere

- The family atmosphere was balmy in that Mother and Father shared the same values, and in that Mother was willing to defer to Father's position as the undisputed leader of the family; it was therefore consistent and predictable in its encouragement of the children in their strivings to achieve whatever they could.

- The family atmosphere was idyllic for a brief period in Paul's early life when the family enjoyed the paradise of their own home. Following the move to the household of the maternal grandparents, the dominant pattern was one of stormy contention, resentment, and complaint, relieved only by humid stretches of disappointment and depression.

- The family atmosphere was tense and explosive, though without any threat of a breakup or a loss of place.

- The atmosphere in this Irish-American family was undisturbed by conflict but equally unrelieved by any sunniness of warmth or affection.

- The family atmosphere was electrically charged, primarily and positively by Mother's social ambitions; secondarily and negatively by Father's failure to meet Mother's expectations and to support Mother's strivings for herself and her children.

- The atmosphere in this upwardly mobile, nomadic, "corporate" family was tense with striving, though relieved by maternal efforts to support and encourage those for whom that striving was difficult.

- The atmosphere in this New American Family of Korean immigrants was "enclosed" against the perceived dangers of life, both physical and social, and had about it an internal storminess of anxiety about money.

- The family atmosphere was both stable and tense until Isabelle was seven years old, when the tension gained the upper hand in the aftermath of Father's loss of employment, the birth of a sickly baby, and Mother's illness.

- The atmosphere in this Scottish-American household was clear and crisp, dominated by the sense of order, of principle, and of propriety.

- The atmosphere in this Cuban-American family was poisoned by the tragedy of Father's accident when Xavier was two years old, remaining tainted by Father's subsequent impairment and Mother's fatalistic resignation, self-denial, and self-abnegation in the role of nurse.

- The atmosphere in this Italian-American family was unfailingly salubrious by reason of the unshakable affection, mutual respect, and confidence which the parents showed to one another, and by reason of the family's solid integration in the life of the church, the neighborhood, and the Italian community.

- The family atmosphere was one of subdued, restrained energy, expressing the tensions of a traditional "inside-outside" family, the inside belonging to Mother and the outside to Father. These tensions were exaggerated by another, less easily understood split: the division between the private life and the public face.

Topic 3. FAMILY VALUES

> 1. Note family values (that is, those values shared by Mother and Father); or, in the absence of either or both parents, those values shared by caretakers.

Case Examples: Family Values

- The family values emphasized by Mother and Father included a genteel anti-materialism; devotion to fair play and equity; and education, not so much for the sake of advancement as for the sake of a kind of moral enrichment.

- The family values emphasized by Mother and Father were: learning, manners, order, loyalty, and responsibility.

- The family values emphasized by Mother and Father were: aspirations for self-betterment, education, interests, and activity. This was a family in which being special or appearing special had a negative value.

- The family values emphasized by Mother and Father were: hard work and congeniality, and a confidence that together these would result in improvement.

- The family values emphasized by Mother and Father were: appearances, timidity, and a certainty that those outside the family (employers, teachers, neighbors) knew better about what was "right."

- The family values emphasized by Mother and Father were: energy, an insistence upon conformity both inside and outside the family, hard work, and affection.

- The family values emphasized by Mother and Father were: orderliness, warmth, concern for others, and the practice of encouragement.

- The family values emphasized by Mother and Father were unconvincing. Each stated the value of honesty, but Tracy was aware of Father's "deal-making" and Mother's secret money cache. Each stated the value of hard work, but Tracy was aware of Father's malingering and Mother's failure to maintain household responsibilities. Each stated the value of kindness, but Tracy was aware of Father's bullying and Mother's harsh judgment.

- The family values emphasized by Mother and Father were those of closeness and harmony. These were exaggerated to the point of constriction, with the consequence of alienating Dean from a sense of participation in the social life of the neighborhood and the school.

- The family values emphasized by Mother and Father included the elevation of the women in the family to imperial positions, with the men playing the roles of courtiers. Mother and Father agreed that Father should dote on Mother and on their daughter, Pam; and that Father's task in life was to ensure that the lives of these two women were comfortable and untroubled by any need to strive.

Topic 4. MASCULINE GUIDING LINE

1. State the masculine guiding line (characteristics, attitudes, interests, and competencies ascribed to Father, and those of Father's values not shared by Mother).

2. If there were other men in the extended family who impressed the client for good or ill, state how their examples confirmed, enhanced, undermined, or contrasted with the masculine guiding line.

Case Examples: Masculine Guiding Line

- The masculine guiding line set by Father stressed strength and strictness without being overbearing or tyrannical. Father was a good-humored, caring, protective man who was respectful of women, so that none of the girls in this family of daughters felt that they were less by being females.

- The masculine guiding line set by Father stressed kindness, friendliness, and, because of his history of heart trouble, a determination to enjoy life fully while it lasted. The guiding line was enhanced by the paternal grandfather, the patriarch whose hard work and whose undisputed authority throughout the family held everyone together.

- The masculine guiding line set by Father stressed competence, intelligence, strength, reverence for the religious life, and an unwillingness to be defeated by disappointment or reproach. As little as Tony knew Father, owing to Father's having left the family when Tony was six, he knew of Father's reliability because Father continued to stay in touch with Tony. The guiding line was supported by the uncles, who presented a picture of religious devotion and generosity toward others.

- The masculine guiding line set by Father stressed a cheerful and friendly interest in life and a confidence in his own standards and in his craft as a journeyman. Father was less interested in appearances than in good relationships and in what he could do. He was also a playful and affectionate man, especially toward his children.

- The masculine guiding line set by Father was marked by intensity, ambition, self-absorption, and impatience with those who did not live according to his specifications. Father could also be petulant and demanding of service, especially of Mother and the girls; the boys were apparently exempted in this regard.

- The masculine guiding line set by Father stressed a restless determination to demonstrate power inside the family, by having temper fits; and to demonstrate acceptability outside the family, by ingratiating himself with others. Father was unconvincing in both efforts, as far as Ray was concerned, since both his power tactics and his "generosity" looked to Ray like weaknesses and posturing.

- The masculine guiding line set by Father stressed a kind of frantic pretense to superiority which expressed itself through sarcasm, faultfinding, belittling, and disproportionate rage reactions (which Wayne saw as "crazy") in the face of any frustration. The overriding difficulty of Wayne's life was his father's disturbed and exaggerated notion of masculine superiority, which Father clung to as if it were a man's duty to dominate and to demonstrate his ability to rule, either by force of argument or by brute force.

- The masculine guiding line set by Father stressed kindness, regularity, orderliness, propriety, and a just and loving concern for those in his care.

Topic 5. FEMININE GUIDING LINE

1. State the feminine guiding line (characteristics, attitudes, interests, and competencies ascribed to Mother, and those of Mother's values not shared by Father).

2. If there were other females in the extended family who impressed the client for good or ill, state how their examples confirmed, enhanced, undermined, or contrasted with the feminine guiding line.

Case Examples: Feminine Guiding Line

- The feminine guiding line set by Mother stressed a kind of relentless goodness, and a determination to apply her higher standards in so conscientious a way as to make her positions seem unassailable to everyone around her. These positions included a perfectionistic striving toward success and the prestige that would come with success, and the maintenance of those appearances that were the signs of that striving.

- The feminine guiding line set by Mother stressed a hard-working attitude of devotion to the family, and a kind of fretful and anxious excitability in the face of any difficulty or any possibility of her being either disregarded or intruded upon. Mother was also capable of another kind of excitement, expressed in gaiety and in an ability to cheer Father up when he was discouraged.

- The feminine guiding line set by Mother stressed a managerial and tutorial style with an unsentimental determination to do what was right in her eyes for everyone else's own good, whether they liked it or not. In the narrow compass of Mother's life there was no room for self-pity or defeat. Mother did not focus on what it was she could not do, only on what she could do, given the situation, to make the best of it.

- The feminine guiding line set by Mother stressed the subordination of enjoyment to the goal of achievement. Mother was also unhappy with her life as a woman, understood femininity in terms of vulnerability, saw herself in danger of humiliation and insult, and also fought against these perceptions by taking a position of superiority to others. Mother's picture of life was that of a person coming from behind (by reason of her Portuguese heritage, which she saw as a disadvantage in the New England "WASP" world she yearned to enter), and it led her into a general discouragement and pessimism.

- The feminine guiding line set by Mother stressed consistency, closeness, and a kind of coercive suffering which was used as a method of control. Mother was a survivor against impossible odds. This made her admirable to Charlie, who felt an obligation to protect her against the threat of any further suffering.

- The feminine guiding line set by Mother stressed fearfulness, weakness, reticence, reclusiveness, and a determination to have her own way and to press other people into her service. Mother was also concerned with appearances and with presenting a brave face to the public world through an air of superiority.

- The feminine guiding line set by Mother stressed an outward pose of submission to the natural course of a woman's life, lived out in caretaking, homemaking, and deference to men; together with an undercurrent of smugness and resentment, as if expressing that, while she would go along, she knew better, and implying a continuous, unspoken grudge against the world.

- The feminine guiding line set by Mother stressed beauty, caring, management, and perfectionism. Mother was selfless in the sense that she pleased everyone, although her hypercritical eye made it impossible for anyone to please her.

Topic 6. THE PLACE MADE BY THE CLIENT IN THE
          FAMILY

    1. Describe how the client made a place in the family.

    2. Note the relationship of the client's place to each of
        the parents and siblings.

    3. Note special alliances and sources of encouragement
        for the client in the family (or address this separately,
        under Topic 8, Other Images).

    4. Note the client's response to the family values,
        including those related to sex/gender.

    Case Examples: The Place Made by the Client

                    The Case of Marsha

    Marsha made a place for herself, by staying ahead of her younger sister by means of her early accomplishments as a bright, responsive child; by being her Father's companion; and by distinguishing herself athletically in the "tennis set" (which was also Father's set), especially by showing that she could handle herself in the company of adults.

She stayed active, took teasing cheerfully (as Father did), and imitated what she saw to be the strengths of each of her parents: Like Father, she knew how to be loud and overwhelming when it suited her purpose; and, like Mother (though she had decided not to grow up to be complaining and reclusive like Mother), she knew how to insist on getting what she wanted through suffering if all else failed.

For the most part, extreme tactics were not necessary to her, and she thrived on the encouraging interest of Father, Grandfather, and Uncle Terry, as well as on the pleasure with which she saw others outside the family responding to her energetic and outgoing manner.

Marsha's response to the family value of education was therefore problematic for her: She was unwilling to subordinate her energy to the fixed and seemingly arbitrary requirements of the school, and set herself on a course of defiance of its restrictions.

The Case of Diane

Diane made a place for herself as Daddy's Little Girl, played to him, capitalized on his willingness to teach, and did her best to do well in school and to be religious and "good," all of which were important to Father.

She did not compete directly with her older brother (perhaps sensing that there was no point in trying to upstage someone who wanted so much to be the Star). She learned to be self-reliant as a way of dealing with feeling excluded from Mother's company (owing to Mother's focus on her brother).

Diane knew that as a tomboy she could do and enjoy all the things that a boy could do. She could not forget, however, that she was not a Real Boy. She was a girl who would be a woman and who, like Mother, would follow the natural course of a woman's destiny. At the same time, she wanted to be valued by a man for those qualities she saw in Father that Mother appeared to her to lack: independence, knowledge, and competence.

In this way she pictured herself succeeding to the position of a woman who could be a fit companion to a man like Father, and her ambition was encouraged and sustained by the examples of her Aunt Ann and Uncle Bill, whom she saw as having achieved a relationship of mutual respect.

Diane's response to the family value of male dominance was therefore to resolve to be a fit ally to a man by being as much like Father and as interested in him and his interests as she could be.

The Case of Murray

Murray made a place for himself in a number of ways. First, he found the place *assigned* to him by virtue of being an illegitimate child, which had both positive and negative connotations: positive, in that he was fussed over by Grandmother as the Only Child of Grandmother's only child; negative, in that his advent had been an embarrassment and difficulty for the family. Second, he was sickly and required a great deal of attention from his caretakers without his having to insist upon it. Third, he was bright, quick to learn, and quick to adapt himself to the requirements of family life with Mother and his grandparents. Both Murray and Mother deferred to the grandparents' attitudes and practices, thereby ensuring harmony in the household and security in the world. Murray took advantage of Grandmother's encouragement to get ahead by doing well in school and by forming good friendships.

The Case of Joyce

Joyce made a place for herself with the only person in her world who could be relied upon: Mother. Mother's on-again, off-again affairs and marriages conveyed to Joyce that it wasn't worthwhile to make an effort to form an attachment with a man. As a consequence she failed to form any close alliances with the men in Mother's life, keeping her attention focused on Mother. She took advantage of Mother's uncertainty and feelings of regret and disappointment, exploited Mother's propensity to feel sorry for her, and so kept Mother in her service. She imitated Mother's competence in the world outside the home, where Mother was a successful professional woman. She also imitated Mother's incompetence in close personal relationships, rehearsing the posture of a person whose life will only "work out" if Lady Luck smiles on her, so failing to cultivate the awareness of her own effectiveness.

The Case of Tom

Tom made a place for himself by following the masculine guiding line set by Father and confirmed by Father's younger brother, Erik, who lived in the household: hard work, intense involvement in his own interests, and a kind of rough determination to rule, expressed by Tom in readiness to beat up on anyone who disturbed his sense of how things should be. Also like Father, he was active and made do with very little in the way of close friendships or close family relationships. The maternal grandparents offered some relief from the rigors of Tom's childhood, Grandfather being an openhearted and openhanded person, and Grandmother being a caring and embracing person.

Because of his restless ambition to do what he could do, and his consequent lack of interest in reflecting upon, pondering, or even attending to matters that were outside or beyond his immediate reach, he began to form a picture of himself as, at best, shallow, and, at worst, dumb. Even so, as the only child of preoccupied parents, Tom understood the family value of independence and action, and set his sights on getting up and out.

The Case of Randy

Randy made a place for himself by being "the little one," alternately overprotected and left on his own by the others. His primary alliance was with his older sister, Lucy, who took him under her wing. Randy had everything he needed inside his family, where he could be little without being bullied or threatened, the center of attention whenever Lucy was around, and free to pursue his own interests whenever he wished. Outside the family it was different, though he was able to make a special place in school by getting along well with his teachers and excelling in academics.

Randy was overly impressed by Father's perceived power, compared to which he saw himself as being relatively ineffective. He determined to make up for this by accomplishments that would take him beyond Father's modest attainments in the world of business. Like Lucy, Mother doted on Randy, comforted him and supported him, and Randy developed a sense that he would be unable to succeed in life without the protection of a devoted woman.

The Case of Margot

Margot made a place for herself by being tough like her brothers, and principled like her father. She admired Mother's beauty, but knew that she looked like Father's side of the family, and so felt that she could not rely upon good looks to get her way as Mother did. She learned to stick to her guns, go after what she wanted with a kind of relentless determination, and to avoid anything that appeared unattainable to her, including the relief of humor and fun, and the comfort to be found in giving and receiving affection.

Topic 7. THE PLACES MADE BY EACH OF THE CLIENT'S SIBLINGS

1. Describe how each of the client's siblings made a place for himself or herself in the family.

2. Note the relationship of each place to that established by the client.

Case Examples: The Places Made by Each of the Siblings
(Clients' names are printed in **bold**.)

- **Bart's** older brother, Alex, made a place for himself as Mother's boy and, because of Father's frequent travels, formed a special alliance with Mother, from which **Bart** felt excluded. Alex also followed the masculine guiding line in being gregarious and in wanting to be powerful, dominant, and knowledgeable.

- **Len's** younger brother, Hank, made a place for himself by accepting Father's values, by pursuing stated goals, and by imitating Father's mannerisms, especially in his way of joking. **Len** felt relieved that Father's attention was focused on Hank, and learned to stay out of Father's way in his own alliance with Mother, who was more easygoing.

- **Frannie's** older sister, Gail, had established a place for herself by becoming Mother's indispensable helper in raising the younger children. This model looked too hard to **Frannie**, who decided to take the position of the inept one who could not be relied upon, and of whom, therefore, less was asked.

- The third child, Larry, made a place for himself by being the Only Boy. His advent was a startling revelation to **Helen**, serving as her first intimation that she had a native handicap in her gender that could not be overcome.

- Margaret made a place for herself by being "offhand," mischievous, socially at ease, and less concerned than **Marilyn** to stick with things. **Marilyn**, as the older of the twins, did what was expected of her and met her obligations, wondering how Margaret "got away with it."

- Charlene had made a place for herself by imitating Mother's propriety and by practicing the religiosity of Mother's side of the family. In this way she affronted Father and kept him busy with her by "cultivating" his disapproval. **Trudy**, aware of this difficulty between Charlene and Father, was all the better able to win him over by taking his part and by otherwise being his ally in the family.

- Carlos had made a place for himself as the Good Boy from whom trouble was not expected, and therefore from whom trouble was not tolerated. By the time of **Jaime's** arrival in the family, Carlos was well-rehearsed in this position, leaving **Jaime** a clear field in which to be as troublesome as he pleased, and to be defiant of the parental efforts to subdue him.

- Ellen made a place for herself outside the family circle, which was tightly drawn around **Debbie**. Ellen became a loner, fighting Mother and Father over what seemed to **Debbie** to be curious and irrelevant issues, a judgment which had the effect of elevating **Debbie's** position all the more as the logical, reasonable, and cooperative one of the two girls.

- Betty made a place for herself as The Baby, competed with no one, and invited everyone to be affectionate toward her by taking advantage of her advantageous position among the others. **Ricky** made a place for himself by making a protégé of Betty, thereby sharing in some of Betty's spotlight.

- Herb made a place for himself by capitalizing on his physical agility, developing his athletic skills, being willing to stay in second place to **Michael**, and cultivating Mother's empathic and sensitive way of relating to others. This enabled **Michael** to pursue and excel in the world of academics and to maintain his firstborn position without strife.

- Nancy operated with thirdborn charm, and avoided the appearance of striving for any position other than the one she had as everybody's favorite. **Margie** enjoyed Nancy and made use of her popularity to gain a place in the wider world of the school.

- Barry made a place for himself by imitating and perfecting **Jacob's** tactic of staying out of the way, withdrawing into himself, and becoming inaccessible, while **Jacob** continued striving to escape the pressures laid upon him by Mother and Father by finding activities outside the home in areas that would be above reproach (Boy Scouts, church), and not likely therefore to invite the interference of the parents.

Topic 8. OTHER IMAGES: ROLE MODELS AND
          ALLIANCES

1. Identify childhood role models (both positive and negative) who were important to the client, and describe the images of possibility or danger they represented.

2. Note other alliances made by the client and the contributions they made to the client's development, particularly in terms of any sources of encouragement.

Case Examples: Role models and Alliances

- Other images presented to Darcie included the relationship between her Aunt Maria and Uncle Hugh. Through their cooperation and mutual support in caring for their severely disabled daughter (Darcie's cousin), they encouraged Darcie to see another possibility for the way in which men and women could establish their relationships, in solidarity and respect.

- Other images presented to Jerry were those of Grandfather James and Coach Hammond. Grandfather James was a generous-spirited man who took an interest in Jerry's interests, as well as an interest in the wider community outside the family, providing Jerry with an example of public service to which he aspired. Jerry made an alliance with Coach Hammond, with whom he shared his athletic ambitions, and from whom he received support and encouragement.

- Sandy made an alliance with her Uncle Jack and with an older neighbor, Mrs. Grayson, both of whom enjoyed her sense of humor and high spirits.

- Another image of manhood was presented to Josh by his Uncle Ron, Father's younger brother. Ron was the "black sheep" of the family, whose periodic appearances, whose rude and aggressive attitudes, and whose self-aggrandizing behavior underscored the dangers of veering off the straight and narrow path.

- Other images presented to Pete included his two grandfathers and his Uncle Dwayne, who were all more relaxed about life than Father, and who demonstrated that a man could be *both* good *and* joyous in life. His alliance with his older cousin, Cynthia, provided him with a rehearsal for give and take with the other sex that would prove valuable to him in his adolescent encounters with girls, a preparation he otherwise lacked.

- Another image of how a woman could be was presented by Ruth's Grandmother Sims, who was the center of the extended family and the one whom everyone loved. Ruth aspired to inherit Grandmother's mantle, and set out to demonstrate that she was worthy of it.

- Lee made an alliance with his grammar school teacher, Mrs. Harris, who recognized and encouraged his artistic talent and inspired him to believe in what he could do.

- The images presented to Patrick by his scout master and the parish priest, of men who were both trustworthy and kind, helped him in his struggle against his pessimistic self-estimate as the son of Patrick, Senior, who seemed to Patrick to be both unreliable and selfish.

- The image presented by Father's friend, Danny, seemed to Marty to be the picture of what he did not want to be as a man. Danny's loud voice, demanding ways, and obsession with money and "making deals" were unattractive to him.

Topic 9. THE EXPERIENCE OF CHILDHOOD IN THE NEIGHBORHOOD AND SCHOOL

1. Describe the client's assessment of childhood experience in neighborhood and school, if this is not incorporated under previous topics.

Case Examples: Neighborhood and School Experience

- During childhood, Teddy defined himself as passive because he was smaller, weaker, less sure of himself in the rough and tumble of the neighborhood and school than the other boys and girls seemed to be. He maintained his courage by harboring a secret picture of himself as a tall, strong, confident grown-up, and made a project of looking forward to achieving that reality simply by surviving.

- During childhood, Sarah managed to escape Mother's management by capitalizing on her social talents. Mother took pride in Sally's many friends and activities and in her success in the world of the school.

- During childhood, Millie found her place among the neighborhood boys, playing ball, exploring the woods, and riding bikes. She felt less comfortable in the world of the girls and women.

- During childhood, Douglas explored to the fullest extent possible the pleasures of his active mind and healthy body.

- During childhood, Ned's place as both the baby and the problem child combined to cause trouble for him among the children in the neighborhood and the school, where he found the others impatient with his demands and intolerant of his obstructiveness. He found relief from these un-understood difficulties only at home, where his mother pampered him and his father left him alone. His love of arithmetic indicates an early desire to find an area of life in which he could figure things out for himself and solve problems on his own.

- During childhood, as the youngest of the boys, Todd admired Father and his much older brothers, and dreamt of a heroic future in which he would excel in masculine achievement and masculine protection of others. At the same time he felt unsure of his qualifications for greatness, especially in his confrontations with the taunts and insults of the other children in the neighborhood. School served as a deliverance. His intellectual ability flowered there and he decided that his way to greatness would be by way of great books.

Topic 10.   THE EXPERIENCE OF PUBERTY

> 1. Describe the client's assessment of the experiences associated with the onset of puberty.

Case Examples: The Experience of Puberty

- Jane met puberty with a sense of accomplishment stemming from her earlier successes in childhood, as well as with a sense of relief: She was now a woman, and the number of her days under her mother's scrutiny were dwindling.

- For Celeste, the onset of menstruation was an unwelcome event as it demonstrated to her that whatever else she might do in life she was doomed to be a woman.

- Lyle welcomed the changes that came at puberty as signs of his becoming a man able to take his place as a man among men.

- Upon reaching puberty, Nate was still undecided about whether to go forward in the direction of an adult world that didn't look all that attractive to him, or to entrench himself in his well-rehearsed posture of mischievous child.

- In puberty, Lawrence's admiring picture of his father and uncles, combined with the intellectual ability that resulted in his competitive success in school, allowed him to form an idea that he could achieve whatever would be required of him as a man.

Topic 11.   THE EXPERIENCE OF THE SEXUAL CHALLENGE
            IN ADOLESCENCE

> 1. Describe client's sexual awareness and experience
>   in adolescence.
>
> 2. Describe the circumstances and client's evaluation
>   of sexual initiation.

Case Examples: Adolescence

- During adolescence, Walt experienced the kind of victimization, heart-break, and bitterness that he had seen as characterizing his father's life; and in the wake of his abandonment by the first love of his life felt more and more destined to live out the masculine guiding line set by Father.

- During adolescence, Elsie's knowledge of her cousin Beth's pregnancy, her learning the family secret about her aunt's abandonment by her husband (who had run away with another woman), *and* her awareness of the special status of the men in her family and extended family, led her to conclude that women were vulnerable to men's desires and decisions, and that men were granted a status in the world to which she could never aspire. Elsie had avoided boys in childhood, sticking with a circle of close girl chums. In adolescence she developed intimate friendships with girls, with whom she engaged in sexual experimentation. At seventeen upon leaving home for college she acknowledged her same-sex orientation.

- During adolescence, Lori had many boyfriends, was the center of a socially desirable circle, and enjoyed her femininity. With the admiration and support of her father and brothers she was never undermined by feelings of inferiority or subordination, and looked forward to taking her place with a man as an equal contributor to their common life.

- During adolescence, Becky consoled herself with her girlfriends, feeling that she could not hold her own among the boys. What dates she had were arranged by others, were awkward, and were not repeated. She yearned for the kind of position enjoyed by Mother who was cared for and catered to; however, she also wanted to follow the masculine guiding line of independence set by Father and reinforced by the other men in the family.

- During adolescence, Elizabeth enjoyed being a girl and looked forward to being a woman with a man, provided that the man were a protector and encourager like Father. Her sexual initiation with the man who would be her husband was a welcome event therefore, marking in her mind her emergence as a woman who would enjoy being a woman.

- During adolescence, Arthur had a restless ambition to get on with it. With the exception of the encouragement that one of his high school teachers gave him in his sophomore year, and his subsequent decision to make use of his remaining years in school, the changes that occurred in Arthur's adolescence were changes in the scope of his movement away from the parental home, resistance to any entanglement with the other sex, and, finally, escape into the armed forces.

- During adolescence, a number of ideas about masculinity and femininity combined in Jason's thinking: The first derived from his parents' reminders that they hadn't wanted more children and that, when Mother got pregnant with Jason, they had hoped that at least the baby would be a girl. The second was that, since (unlike his brother) he had not been circumcised, there was "something wrong" with him. Third was his idea that the task of dealing as a

man with a woman was a doomed endeavor: Women were
unreliable, unpredictable, untrustworthy, and irrational. Finally,
there was his idea that a man's life was hard, and that a woman's
life was comparatively easy, exempted as it was from the rigors of
economic struggle outside the home. These things conspired in
Jason's thinking to train him away from the masculine role: It was
too hard, and he was unqualified (uncircumcised). He began to turn
his attention away from the other sex, toward members of his own
sex.

- During adolescence, everything went smoothly for Derek. He
knew what he must do: Play by the book. As long as everyone else
followed the rules, things would proceed as they should. Because
of his energy, his lack of abrasiveness or antagonism, and his good
looks, girls were attracted to him; and because of his enjoyment of
them he had no trouble making friends with them and winning
over the one he wanted. His trouble came later when his girlfriend,
now his wife, wanted to play by a new set of rules, and he was
unable to write a new book.

Topic 12. MAJOR UNRESOLVED ISSUES REMAINING
            FROM CHILDHOOD AND ADOLESCENCE

1. Identify and explicate any issues remaining from
   incomplete training and self-training that confront
   the client at the threshold of adult life, making note
   of any self-imposed limits hampering movement,
   and the conditions under which the client is prepared
   to go forward.

Case Examples: Unresolved Issues

- The question confronting Chris at the threshold of adulthood
remained: How could he combine a sure thing with the main
chance? On the one hand, he would keep things under control
without becoming rigid, inflexible, or odd (like Mother); on the
other hand, he would play at gambling in the markets, using his
mathematical knowledge of the odds to keep himself from making
dumb mistakes (like Father). The consequence of this training in

double safeguards seems to have left him in a position in which he suffers from both dangers. When (like Father) he makes a mistake, he feels stupid; then, when he tries to keep himself from making further mistakes, he paralyzes himself, takes no further action at all, and feels odd (like Mother).

- The question that confronted Carla at the threshold of adulthood was: How could she manage the "wild energies" of the women in her family so that she might bind the beauty and the power of a man to her. Should she, like Grandmother, use these energies to overwhelm and *tame a man*? Or should she, like Mother, *tame herself* and thereby hope to bind the man to her?

- The question Alan confronted at the threshold of adulthood remained: How could he be a man without having to do what men do, namely, accede to the demands of perfectionistic women who claim a right to have their demands met by reason of their suffering?

- The question confronting Kim at the threshold of adulthood remained: How to be the man her father was and still remain eligible for the position of the woman? How could she succeed in the world of men, the world of hardship and accomplishment, and still succeed in the world of women, the world of comfort and indulgence?

- The question that remained to confront William at the threshold of adulthood was: How to escape as Father had escaped? He was unable to be like Father, however, in that he was haunted by the opinions of others that undermined his courage and distracted him from finding for himself, in any situation, a clear course by which to proceed toward his own goals and his own fulfillment.

Analysis of a Family Constellation Summary

Now that we have set forth the topics for a family constellation summary and have illustrated them with excerpts from cases, we can turn to considering how they were applied in a single case. For this we return to "The Case of Dan," presented in Chapter 4. In the material that follows we use Dan's *Summary of the Family Constellation* (pp. 117-120) for a step-by-step analysis, first stating each topic, then, by providing relevant quotations, illustrating how the topic was addressed in the summary.

Bear in mind that because each lifestyle is unique, each application of these topics is also unique, adapted to the one situation of this one person, as you will see below.

The Case of Dan

Topic 1. Birthorder

> 1. Name the client. 2. State the ordinal position. 3. State the client's psychological vantage vis-à-vis siblings.

"Dan grew up as the older of two boys, psychologically the firstborn son."

Topic 2. Family Atmosphere

> 1. Describe the family atmosphere set by the undertone of the relationship between the parents (or other adult caretakers resident in the childhood household). 2. Note any extraordinary changes in the atmosphere and intervening life events relevant to such changes. 3. Note the ethnic/racial character and traditions of the family if significant in establishing the atmosphere.

"The family atmosphere was one of disturbed security. The security lay in knowing that Father would provide for and protect the family; the disturbance lay in knowing that Mother would, nevertheless, be anxious, unhappy, and dissatisfied."

Topic 3. Family Values

> 1. Note family values, that is, those values shared by Mother and Father, or, in the event of the absence of either or both parents, the values shared by caretakers.

"The family values emphasized by Mother and Father included hard work, caretaking of what one has (house, garden), and religious feeling."

Topic 4. Masculine Guiding Line

> 1. State the masculine guiding line (characteristics, attitudes, interests, and competencies ascribed to Father, and those of Father's values not shared by Mother). 2. If there were other males in the extended family who impressed the client, state how their examples confirmed, enhanced, undermined, or contrasted with the masculine guiding line.

"The masculine guiding line set by Father stressed a kind of relaxed cheerfulness and steadiness, together with a balanced temper and a practical intelligence capable of flashes of brilliance. Father was also capable of putting up a show of fighting back, and knew how to pursue his own interests quietly even when opposed (playing the horses). Grandfather confirmed the masculine guiding line in his loyalty to the women in his life. Uncle John enhanced the masculine guiding line through his sense of humor and relative material success."

Topic 5. Feminine Guiding Line

> 1. State the feminine guiding line (characteristics, attitudes, interests, competencies ascribed to Mother, and those of Mother's values not shared by Father). 2. If there were other females in the extended family who impressed the client, state how their examples confirmed, enhanced, undermined, or contrasted with the feminine guiding line.

"The feminine guiding line set by Mother stressed a fretful, ambitious, pessimistic, and temperamentally unpredictable busyness. Mother was better than others, and yearned to be among her betters. Mother's failure to express warmth was confirmed as characteristic of femininity by the attitudes of Dan's Great Aunt and his Aunt Jill, both of whom were cold. Even Aunt Carla, a generous, comfortable woman, kept her distance by 'sending' presents, and by avoiding any alliance with a man."

Topic 6. The Place Made by the Client in the Family

> 1. Describe how the client made a place in the family. 2. Note the relationship of the client's place to each of the parents and siblings. 3. Note special alliances and sources of encouragement for the client in the family (or address this separately, under Topic 8). 4. Note the client's response to the family values, including those related to sex/gender.

"Dan stayed ahead of Douglas by sharing Father's interests, and by being Mother's chief object of interest. Mother, who saw her life in tragic and heroic terms, was determined to see to it that her sons would succeed in ways that had been foreclosed to her by what she regarded as unfair circumstances.

"It was difficult for Dan to understand Mother's fretfulness about appearances, about striving, about education and moving up and out in the world, and about Father and his alleged inferiorities.

"In the world of the home, Dan apparently imitated Father's way of quietly enduring Mother's agitation, which led to his enduring her fussing over him as well. He also imitated Father's temper outbursts, by means of which he (and Father) were able to gain occasional relief from Mother's pressure. In Father he found a source of acceptance, but no standards by which to measure progress. Mother, who praised him when he met her standards and chastised and withdrew from him when he did not, provided the source of judgment."

Topic 7. The Places Made by Each of the Client's Siblings

> 1. Describe how each of the client's siblings made a place for himself or herself in the family. 2. Note the relationship of each place to that established by the client.

"Douglas, three years younger, made a place for himself by leaving the first-place position to Dan, uncontested; by staying out of trouble; and by developing an easygoing, friendly attitude toward life. Seeing that Mother's focus on Dan was the source of Dan's troubles, Douglas contrived to stay out of Mother's way, and out of her 'spotlight.'"

Topic 8. Role Models and Alliances

> 1. Identify childhood role models (both positive and negative) who were important to the client, and describe the images of possibility or danger they provided. 2. Note other alliances made by the client and the contributions they made to the client's development, particularly in terms of any sources of encouragement.

"Dan found another source of acceptance in his Uncle John, whom he often saw in Father's company. John encouraged Dan, took pleasure in the boy's efforts, recognized his intelligence, and, like Father, did so without judgment."

Topic 9. The Experience of Childhood in the Neighborhood and School

> 1. Describe the client's assessment of childhood experience in neighborhood and school, if this is not incorporated under previous topics.

"In the world of the neighborhood, he was able to do well enough without striving, buttressed by a sense of being 'a cut above' the others that carried him through his difficulties with them. His good appearances, maintained by Mother, and the good appearance and steadiness of the household, probably sustained him in this idea."

Topic 10. The Experience of Puberty

> 1. Describe the client's assessment of the experiences associated with the onset of puberty.

"He managed well enough in the world of the school until adolescence, when everything changed: He now found himself propelled into a new group of peers, and compelled to assess his personal situation against the background of a much wider social world."

Topic 11. The Experience of the Sexual Challenge in Adolescence

> 1. Describe the client's sexual awareness and experience in adolescence. 2. Describe the circumstances and the client's evaluation of sexual initiation.

"Dan suddenly saw himself as a very small duck in a very large pond. His uncertainty about his place in the larger world, beginning at this time, continued to express itself in his hesitation to engage with someone of the other sex in caring intimacy, until in his mid-twenties he could make his first effort to do so under the protection of a formal engagement."

Topic 12. Major Unresolved Issues Remaining from Childhood
            and Adolescence

> 1. Identify and explicate any issues remaining from incomplete childhood training and self-training that confront the client at the threshold of adult life, any self-imposed limits hampering movement, and the conditions under which the client is prepared to go forward.

"In adolescence, all of Mother's complaints against Father's lack of ambition, and all of Mother's warnings against Dan's being 'like the Allens,' previously hard for him to understand, came to make a certain kind of sense to him: appearances in his new world were much more important than he had thought. It was not even enough to look good; he would have to sound good as well, with the proper accent. He came to see that Mother's brand of ambition and striving were essential if he wanted to better himself. (Unlike Father, he could not afford to miss any buses; for him, another might not come along.)

"The question that confronted Dan at the threshold of adult manhood remained: How could he prove himself to be a man unlike the Allen men? To do this, he might suppose, would require that he set out in quest of an unhappy woman, and then demonstrate that he could make her happy by becoming whatever she wanted."

Conclusion

Chapter 13 has presented a method for constructing the family constellation summary, the first of the lifestyle assessment summaries. In our practice, this summary is dictated in the presence of the client after a review of the relevant material. Other professional settings may require other ways to compose and present the summaries.

In the next chapter we discuss our method for examining the client's early recollections (1) to summarize the client's pattern of basic convictions, and (2) to isolate from among the convictions those we identify as the client's mistaken ideas.

Exercise for Chapter 13.
Summary of the Family Constellation
(Use in dyads, small groups, or class, or with clients.)

Use the format below (p. 388) to organize lifestyle data. For further assistance, review pp. 354-355, Topics 1 through 12, and the case illustrations that follow. If parents were missing in client's childhood, use client's acknowledged parental substitutes or client's image of a missing parent, even if fragmentary. (If there were no siblings, skip Topic 7.)

Topic 1. [Client's first name] grew up as the [state client's ordinal position], psychologically the [enter the client's *psychological vantage*]. . . .

Topic 2. The family atmosphere was. . . .

Topic 3. The family values emphasized by both Mother and Father stressed. . . .

Topic 4. The masculine guiding line set by Father stressed. . . .

Topic 5. The feminine guiding line set by Mother stressed. . . .

Topic 6. [Client] made a place for herself/himself by. . . .

Topic 7. [Sibling's name] was [state age and position relative to client], made a place for herself/himself by. . . . [continue for each sibling.]

Topic 8. Other images presented to [client] were [identify role models, both positive and negative, describing their significance to the client]. [Client] made an alliance with [identify the person] who was important to [client] because. . . .

Topic 9. During childhood [client's] experience was. . . .

Topic 10. Upon reaching puberty, [client's] experience was. . . .

Topic 11. During adolescence, [client] felt . . . and met the challenges of self-reliance, sexuality, and direction for life by/with. . . .

Topic 12. The question confronting [client] at the threshold of adulthood remained [here identify client's lack of preparation for adult success; include conditions under which client is prepared to go forward].

Chapter 14. THE PATTERN OF BASIC CONVICTIONS AND THE MISTAKEN IDEAS

Other Individual Psychologists use other names for the two summaries that conclude the lifestyle assessment. Dreikurs called *The Pattern of Basic Convictions* "The Summary of the Early Recollections," which is an accurate way to describe the procedure itself. We prefer our name because we believe that it has more meaning for the client. We use the word pattern in its double sense: as a design to show how each of a person's ideas is related in the whole style of the person's movement; and as a guide by which something is made, as a carpenter's or dressmaker's pattern, and in this case, as the underlying guide for the person's movement.

Similarly, some of our colleagues have used the terms "basic mistakes," "fundamental errors," or "erroneous convictions," and elsewhere we have used "interfering ideas" for what we now prefer to characterize as "mistaken ideas." We prefer our name "mistaken ideas" because it presents the client with an awareness of the trouble these ideas introduce, as an added cost in the business of life. Because it does not hint at anything more dire or durable or basic to deal with than ideas, it seems to us to be a more encouraging term. Whatever the nomenclature, the two summaries are based on an understanding of the same data, namely, the core sample of the lifestyle found in the client's early recollections.

Reviewing the Early Recollections for the Summaries

The therapist must identify this core sample in such a way as to display it as a model of the client's movement, and with such an accuracy and lucidity (as Adler's statement introducing Part IV prescribes) as will let the person feel both understood and able to recognize his or her mistakes. This is the heart of the psychoclarity process.

The Case of Janice

You will remember that in Chapter 11, after completing the initial report and discussion of Janice's early recollections (pp. 294-320), a brief review took place. In this review, the task of the recording therapist was to remind the consulting therapist (whose task it would be to compose the summaries) of each early recollection in the record, with the client's age at the time of the event recollected. During this review the early recollections that had remained unclear (as to their individual significance and meaning, or as to their place in the record) were discussed further with the client. (For example, the Santa Claus memory and the rooftop memory required further exploration before their meanings came clear.)

To facilitate the fixing of the early recollections in mind, the consulting therapist may choose to write down a list of the client's age and key words or phrases. Following is the way these notes were written down by the consulting therapist in the course of the discussion with Janice:

   4 - Car. Observe (listen). Safe. Good.

   7 or 8 - Snake. Blunder. Energy. Fear.

   8 or 9 - Basement. Water/dark/still. G-friend. Peaceful. Good.

   4 or 5 - Santa. Strength/necessity. Yield.

   6 or 7 - Intercourse. Dark. Sounds (uninhibited). Afraid.

      8 - Thanksgiving. FA takes pic with g-friend. Pretty. Good.

      9 - Halloween. Ghost. Hidden/free. Good.

      9 - Roof/between boys. Woman below. "Hanging on."

Factors to Consider in Composing
The Pattern of Basic Convictions

Working directly with Janice's early recollections, we can illustrate a method for forming this summary by studying the data with reference to three major factors: *sequence, similarity, symmetry.*

Sequence means the arrangement of the entire set of early recollections, and the relationship of each to the ones before and after it in the series.

Similarity means a repetition of (or a variation upon) ideas or themes in various early recollections. An example is the idea of concealment appearing in various forms in Janice's stories.

Symmetry means correspondence between two early recollections as they balance one another, or are juxtaposed as if to represent two sides of an issue. For example, if one early recollection addresses the relationship between the client and women, and another between the client and men, we have symmetrical early recollections which should be reflected in the summary by: "Men are. . . . Women are. . . ." Likewise, one memory may tell what happens when I act on my own initiative, and another what happens when others lead the way or require things of me. One may illustrate life as it should be (or, as it could be if only. . .), and another may illustrate life as it is, perhaps in a contrast of paradise and paradise lost.

We can now examine Janice's early recollections with respect to these three factors of sequence, similarity, and symmetry. We do this in an informal conversation, with the client participating in the process.

Sequence

To identify sequence, make a précis of each recollection, moving from one to the next. Here is a list of titles for Janice's recollections, followed by a series of numbered paragraphs showing how the consulting therapist recapitulated the discussion of each recollection (see Chapter 11) in identifying the sequence:

> Car
> Snake
> Basement
> Santa Claus
> Intercourse
> Thanksgiving
> Halloween
> Roof

1. We go from the car, where I am hidden in the dark, an observer, where others are in charge, and where all is contained and safe, to

2. being on my own, still in the dark, and blundering, and letting loose an energy that is uninhibited, and associated with *the masculine* (belonging to my brother), and it's frightening, to

3. the basement, again dark, again hidden, where I'm with my girlfriend, contemplating the containment and peace of the pool of water, associated with *the feminine*, to

4. a world ruled by a man who is larger than life, who is in charge, and who can force me to do something I don't want to do, and even though I struggle I finally have to yield, and when I do yield it's all right, and he promises to bring me presents, whatever I want, to

5. a picture of how this works between a grown woman and a man, where it's dark, where she is hidden, and she is struggling, but he's bigger and stronger and on top, and where these

strange noises express an untamed energy, and it's frightening, to

6. being on the lawn on Thanksgiving, again with my girlfriend, looking pretty, contained in our dress-up clothes, and being admired in the open light of day by the man, from afar, and I am happy to be admired by him, and to pose for him, to

7. being hidden from everyone in my ghost costume, enjoying the dark, even though I know and they know that I am here, enjoying the charade we're carrying out, to

8. being hidden again, this time in a perilous situation, though I know that even if I let go and fall from the roof I won't be too badly hurt, and it's just that I don't want to, I don't want to give up my position as one of the kids and take my place down among the women and have to do what they do, and I know that I can't hold on forever, but I'd rather pretend for as long as I can, even though they know and I know that it's a game.

Similarity

The next task is to identify similarities, such as common themes and the repetition or restatement of ideas. Here is the list of memories again, with our notes for each item. Question marks in parentheses (?) indicate hypotheses in formation:

Car: Containment. Dark. Hidden. Concealed.   Observer.

Snake: Dark. Snake hidden, concealed until I release it.
         Energies are uncontained. Masculinity =
         "uncontainable," "indomitable." (?)

Basement: Containment. Dark. Observer. "Below stairs" in
                 the world of women (?) I'm there with my
                 girlfriend. Femininity = "contained," "subdued."
                 (?)

Santa Claus: Daylight. "Upstairs" in the world of men in the
                 department store. Can struggle, but must yield.
                 Then, promise of reward. (?)

Intercourse: Dark. Mother hidden, struggling. Uninhibited
                 sounds = uncontained energies. (?)

Thanksgiving: Daylight. Man's world. He admires, we pose.
                 He's at a distance. I'm with my girlfriend. There
                 is a woman's world and a man's world. (?)

Halloween: Dark. Hidden. Concealed. Observer. Awareness
                 of its being a game. No winners or losers; only
                 participants.

Roof: Daylight. Hidden, up in the man's world.
                 Concealed. Again, an observer. Again, aware of
                 its being a game. If I "fall" the important thing is
                 not that I'd be hurt; it's only that I want to hold on
                 and hide for as long as I can before I settle
                 for/settle down to my place as a woman. (?)

Symmetry

With sequence and similarity in mind, the next step is to look for
symmetry, that is, those juxtaposed stories that illustrate the
antithetical frame of reference. In Janice's case, we found the
following:

Containment vs. Absence of limits or form ("wildness"). The car
and pool memories, and the snake and intercourse memories
provide symmetry.

Hidden vs. Exposed. I can be hidden by taking my place apart, in the child's position (car); below, in the woman's world (basement); by my secret actions in the dark (snake); and by my disguising myself so that no one knows whether I'm a man or a woman (ghost). Either that, or I am exposed in the daylight of the man's world, where my subordination and weakness can be seen (submitting to Santa Claus, posing for the photo).

The man's world vs. The woman's world. The women are below, along for the ride, objects of masculine admiration and desire, and receive presents in exchange for submission (sex), constraint (ornamentation), and service (laundry). The men are above, in charge, wildly exciting, unyielding, frightening, and (if the woman gives in) benign.

The game of life vs. Engagement with life. It's exciting to stay in hiding (car), to keep myself in the dark (basement), and even to make believe that nobody knows what I'm up to (Halloween). Anything else threatens to expose me, a pretender (roof) or a blunderer (snake), as posing (Thanksgiving photo).

Dictating *The Pattern of Basic Convictions*

Now, on the basis of the original review and interpretations, the second brief review, and an examination of the entire set of early recollections with reference to sequence, similarity, and symmetry, we are ready to dictate a summary of Janice's basic convictions. Notice that this summary is a first person singular statement, as if Janice herself were speaking. It is dictated in her presence.

The Case of Janice

The Pattern of Basic Convictions

I am part of things and apart from things, happy to be carried along in the stream of life, and able to observe and to listen from the security of my privacy.

When I try to act on my own, I blunder into the discovery of the energies that only men know how to handle. These energies are fascinating but frightening, because they seem so uncontrollable to me.

Feminine things are still and mysterious. We women are in on them together, without having to talk about it. It doesn't help for a woman to resist the imperatives of masculine power, however, and I know that my peace comes through yielding to it, however strange it may seem to me at first.

All the same, there is something disturbing about the inevitability of the submission that is required of a woman in her life with a man. I make a show of resistance to it, and I pretend that I can take my place as "one of the boys."

It is easier for me to be a woman with a man when he also keeps his distance and doesn't try to see beyond my appearances. I am happy that my soul is my own, and that I can keep my desires hidden from the scrutiny of others.

I only wonder how much longer I can hold on to my position and maintain my pretenses, before I am discovered and exposed as a woman who has overreached and disgraced herself by trying to act like a man.

Client Reactions

After the dictation and recording of *The Pattern of Basic Convictions*, the therapists ask the client for any reactions to the work. Having taken part in the entire discussion of the early recollections, and having assisted in formulating the conclusions reached, the client is usually in accord with the summary, even though the therapists may need to clarify some phrases or make some stylistic changes.

Frequently the client is startled at hearing the dictation. Having one's own most fundamental and unconsciously held attitudes and convictions put into a focused and coherent statement is a powerful (and uncommon) experience of being understood. If we attain to our goal of "a real explanation must be so clear that the patient knows and feels his own experience instantly" (Adler, p. 335), we may expect to see some form of the recognition reflex: Clients may respond with expressions of profound relief, sometimes in a sigh, a sucking in of the breath, an exclamation, or a free, unembarrassed flow of tears. It is wise to allow time for this. An experience this far out of the ordinary can be disorienting; the client's efforts to maintain composure, if there is no invitation to react freely, will block the ability to attend to anything more.

Factors to Consider in Composing
The Mistaken Ideas

We want to establish the context [of the error] and make it clear to the person. We want to convince him to the point where he cannot take a further step without this conviction [that this is an error]. . . . If he actually recognizes his error — if he understands the connection and persists in his attitude despite the harmfulness involved — we can only say that he has not understood everything. I have not yet seen a case of this kind. *Really* to recognize an error and then not to modify it runs counter to human nature; it is opposed to the principle of the preservation of life. (Adler, 1963, p. 3)

After the brief exchange inviting, allowing, and responding to the client's feelings, the therapist states the final summary, called *The Mistaken Ideas*. This can be introduced with a comment such as, "Now let's see if we can isolate a few of the ideas that are likely to be contributing to your difficulties. Ready? Number One. . . ."

What follows is an unusual procedure. Everyone knows how discouraging it is to be criticized, and how little we are prepared to learn from people who assault us with statements beginning, "The trouble with you is. . . ." How then can the therapist hope to identify the client's errors in a convincing, and encouraging, way? How can we say, "Some of your ideas — the ones that appear to be expressed in the movement of your life until now, as revealed in your early recollections — are mistaken"? How can we know that we are not mistaken when we say it?

The answer lies with the client, with whom we are not engaged in an argument to discover which of us is right, but with whom we are working to find a better understanding. We are not out to gain assent to *our* position; we want to awaken a recognition of the errors in the client, who can then see that there is a way open to a new position.

We try to remember that the errors are in the person's movement, in his or her "step" or "attitude," to take Adler's terms from the paragraph quoted above. The ideas we are referring to here are not in the client's thoughts somewhere, conscious or unconscious (although they may be logically implied by some of those thoughts).

The mistaken ideas are therefore more properly understood as artifacts created by the therapist and introduced into the client's thoughts, to bring the meaning of the client's movement into clear awareness.

We remind the reader of "Edward," in the first case presented in this book (Chapter 2, pp. 20-26), and our observations there about the arrangement made to safeguard self-esteem. Our argument was that the *private sense*, being antisocial, cannot be put directly into words and cannot be expressed clearly in language because language itself is a common creation, intended for the expression of a *commonsense*.

Private sense and language are therefore incompatible; to keep a private sense of movement from being exposed as antisocial, it must be kept unspeakable and so, also, unthinkable, in the pain and constriction of the psychomuddle we spoke of toward the beginning of this book (p. 29).

To state the mistaken ideas is, therefore, to confound the logic of the arrangement, and to speak *as if* the private sense could be stated in the common language. It is to frame statements that would have to be true (and clearly, to the common sense, cannot be) for a person's behavior to be intelligent and appropriate. Now, in the course of the psychoclarity process, and with the client's collaboration, we have introduced these statements into our conversation in such a way as to reveal their impossibility and absurdity.

Discouraged people are innocent of the implications of their quests for positions of personal superiority. In one way or another, it is as if we are confronting the client's demand that, in order for him or her to be free to fulfill a private goal of personal superiority, everyone else is required to be satisfied with inferior positions by comparison. We are speaking about the unspeakable; the unthinkable can now be thought of for what it is.

The therapist is, of course, a fellow human being, also with a schema of apperception that is not without biases. This is why it is absolutely essential for us as therapists to be aware, as much as possible, of our own recurring biases and the errors to which we are prone. Everyone who works as a therapist should have had the experience of being in therapy, and everyone who aspires to assess the lifestyles of others should collaborate with someone already experienced in this work for a thorough, convincing, and detailed lifestyle assessment of his or her own.

Furthermore, no one should work alone with a client. The intensity and intimacy of understanding that can be achieved through a successful lifestyle assessment occasions opportunities for self-aggrandizing behavior on both sides of the transaction. Every therapist works better, more honestly, and more clearly for the client's benefit when there is a co-therapist sharing the task. Failing the availability of a co-therapist, we recommend that you seek out a senior colleague for regular case consultation.

Assuming that these conditions are met, we must still consider the theoretical question of how one error-prone human being can identify errors in the movement of another. Individual Psychology approaches this task by asking where there is a departure from *social usefulness* in the particular conviction that would have to be correct for the person's movement to make sense.

Gemeinschaftsgefühl (G.), Adler's concept of the community feeling, provides the frame of reference. This term translated most accurately as community feeling (though most commonly as social interest), refers to the individual's awareness of belonging to the human community and the extent of his or her sense of being a fellow being. This feeling can be thought of as an index to successful adaptation: The more developed the community feeling, the more diminished the individual's feelings of inferiority, alienation, and isolation.

Gemeinschaftsgefühl is, therefore, as Adler said, "more than a feeling; it is an evaluative attitude toward life" (p. 135). Because of the importance of this concept, we will take some time here to expand upon it.

Community feeling is the awareness of being one among others with an equal share of responsibility for shaping the common life. It has these characteristics:

1. It is an inherent potential of human life.

2. It is culturally and individually developed, as the capacity for speech is developed in the course of acquiring a language.

3. It is an evaluation of self, others, and the whole of life, indicating the extent of interest in the human community, its inheritance from the past, its present situation, and its survival into the future.

4. It is manifested in cooperation and in identification with others.

5. In the development of our common life it inspires conformity to useful custom, with good humor; resistance to injustice, with courage; and rebellion against exploitation, with resolve.

6. It is therefore an index to successful adaptation.

The Mistaken Ideas is a summary statement of convictions that fail to fit the requirements of life, but would have to be present to account for and make a client's misdirected movement intelligent — and intelligible.

In other words, these mistaken ideas answer the therapist's question, *"What would have to be true for this client's movement to make sense?"*

We can identify these distorted private truths as biases in the antithetical mode of apperception. These are the exaggerated and exclusive either/or ideas in a person's apperceptive schema that correspond to "the primitive attempts of the child to orient himself in the world and to safeguard himself" (Adler, p. 249).

Antithetical Categories

We provide a list of antithetical categories in Figure 8 (p. 403) to exemplify those *either/or* convictions that serve as the foundation of erroneous movement. These convictions are not to be thought of as polarities at one extreme or the other on a continuum. Rather, they are to be thought of as *mutually exclusive* beliefs, operating as "heads or tails," in opposition to each other, circumscribing possibility.

As no one can move in two directions at once, the fictions of antithetical thinking are seductive: They simplify choice. They are useful in practical matters, such as "Either I turn left or I turn right." But they are mistaken, even disabling, when invoked as imperatives in matters of value: "Either I must be faultless or I will be worthless." Beyond oneself, antithetical thinking may enter into evaluations of others and of the world. In the tyranny of fanaticism, for example: "Either I submit to dictated thought or I lose my right to be heard and perhaps my right to live."

For purposes of composing *The Mistaken Ideas*, antithetical convictions may be framed this way: "*Either* I am/others are/the world is _____, *or* I am/others are/the world is_____."

Anything *symbolic* of loss of status extends to the idea of death. Expressions such as "I'd rather die than have to do that!" "He's murdering me with his sarcasm!" "She'll kill me if she finds out!" are more than hyperbole. The social embeddedness of individual life is complete — there is no life for us that does not depend on our capacity to live among others. Our status among others can therefore be experienced as a matter of ultimate significance.

In every case of antithetical reasoning, fictions relating to security, power, and superiority (often including masculinity) are attached to one side of the antithesis, while total loss of these things is attached to the other.

At the head of this list stands (either) masculine (or) feminine, perhaps the clearest example of troublesome oppositions. The mistaken opposition in the other examples offered in Figure 8 is of the same kind.

The antitheses show the ways in which the great line of life, the line of striving from below to above, is traced — or missed — in an endless variety of individual fictions. In enumerating the mistaken ideas we are asking our clients to recognize that, in creating the unique fictions on which their styles rest, they have gone beyond what is sensible and socially useful. They have imposed requirements on their lives that life does not impose; they have set terms and conditions upon their readiness to live among others that the others are unable to meet.

Categories of Antithetical (Either/Or) Thinking

Either	**Or**	**Either**	**Or**
masculine	feminine	selfless	selfish
above	below	adored	despised
good	bad	uncommitted	trapped
right	wrong	aggressive	malleable
perfect	defective	cynical	gullible
heroic	cowardly	assertive	passive
competent	blundering	faultless	worthless
indulged	rejected	rigid	flighty
duty bound	defaulting	rational	obstinate
impressive	inconsequential	glamorous	unpresentable
respected	scorned	scrupulous	irresponsible
dominant	submissive	just	despotic

Figure 8.

The Case of Janice

Composing the Mistaken Ideas Summary

In Janice's movement, as revealed in her early recollections, we noted exaggerated ideas about the constituent characters of masculinity and femininity; about social positions above and below; about the presumed danger of being exposed; and about fears of blundering and letting loose "forces" she is unable to control.

It was our task to put these ideas into clear statements of fact, in common language from the standpoint of the commonsense. To do this requires a certain artistic abstraction of the ideas from the muddle of their unarticulated place in her images. We want to show her that those images imply impossibilities, that they are, at best, caricatures of the truth.

One more thing is required, if we are to succeed in this: We must, to borrow a homely metaphor Adler once employed, "besmirch a clean conscience." It is not enough to show her how she suffers as a result of her errors. She knows better than we that this is true; this is what brought her to see us. Of course she wants to be relieved of pain. What she cannot see is the place that pain plays in an arrangement she has created, and the trouble it causes for others. In every departure from the genuine requirements of cooperative social living there is an antisocial element (review Figure 2, p. 17). It is this that needs to be exposed, so that the innocence of her suffering can be replaced by a responsibility for her share in shaping our common life.

The Case of Janice (continued)

The Mistaken Ideas

1. She has an exaggerated idea about the value or the importance of keeping her thoughts and feelings hidden and private, as if she were fearful of taking responsibility for her own life.

2. Her ideas about masculine and feminine are confused with exaggerated notions about energy and passivity. Men act; women endure. Men dominate; women are subdued. Therefore, men are subject to criticism; women are free to criticize.

3. She has some idea that she is capable of acting recklessly and thereby of provoking or loosing forces, presumably masculine, beyond her control. She therefore believes that she cannot be straightforward in her dealings with men, and feels safe only when she is striking poses for them.

4. As a further consequence of her notions about masculine and feminine, she has some idea that she has succeeded in "rising above" her proper destiny as a woman by means of her professional achievements, and she has a matching apprehension that she won't be able to hang on to this position. In this way she privately transforms her accomplishments into superhuman (or at least super-feminine!) heroism.

After *The Mistaken Ideas* are recorded in the presence of the client, we again ask for a reaction to the work. Assent to the ideas is almost always expressed; opposition to any of the conclusions is rare. Sometimes a client will say, "I don't get it." We have to be prepared to respond, "We may be wrong. This is only the way it looks to us on the basis of everything we've talked about so far." Unless we can find a word or phrase that is more readily recognized by the client, we ask only that he or she think it over until our next meeting, with an assurance that we have recorded nothing more than a working hypothesis for further discussion.

Concluding the Lifestyle Assessment

With the recording of *The Mistaken Ideas*, we complete another phase of the lifestyle assessment. In our practice, there are two more steps: a session with one of the therapists devoted to reviewing the typescripts of the summaries of the three parts of the assessment (Initial Interview, Family Constellation, and Early Recollections), followed by a final evaluation session with both therapists.

After the summaries are prepared, the client and recording therapist meet to discuss them. They first address the client's thoughts and feelings following out of the previous session when the early recollections summaries were presented. The client is then asked to read each summary aloud, one statement at a time, pausing at any point to react (for example, by asking for explanations of ideas that are not understood, by raising questions, or by affirming recognition and agreement).

Reading the summaries aloud is an important exercise. It moves the new way of looking at things one step further from being an imposition of the therapist toward being a work wrought by the client. With each assent to the statements made in the summaries, the client is acknowledging a new orientation as authentic and claiming the story as autobiography. Each difficulty the client has in recognizing what is meant by a statement opens the way to further discussion, to reminders of what was said in earlier meetings, and to a finer and more precise focus.

During the reading and discussion the therapist makes notes of issues raised, adding them as footnotes to a copy of the transcript. The client retains the original.

The final session of the lifestyle assessment includes the client and both therapists. It is the time to review and consolidate learning, to critique the experience, and to make plans for future work together, if that is called for.

In the final session the client is invited to report on further reactions to the experience of reviewing the summaries. The recording therapist then presents the notes taken during the reading of the summaries. Any issues that may have come up since are discussed. (These will sometimes relate to the client's having shown the summaries to family members, who may have added information or raised new questions.)

After this discussion, the client, reminded that this is the last session in the lifestyle assessment, is asked to reflect on and to evaluate the experience and its outcomes. Finally, in the light of this evaluation, the client and therapists weigh the options for further therapy. Sometimes, as Dreikurs (1973) observed, completion of the lifestyle assessment brings therapy to an end:

> There are patients who are so shaken by the mere disclosure of their "arrangements," by the recognition of their accustomed goals, by the discovery of their [own] control of their future deviation and mistakes, that this experience alone furnishes the momentum for successful conclusion of the cure. (p. 24)

In other cases, the client elects to build upon what has been accomplished, and in this final session decides which of the two therapists to work with as the principal counselor, and which will function as consultant, who will attend sessions from time to time (typically, each fourth or fifth session) in order to maintain the advantages of multiple therapy.

Conclusion

What happens in the psychoclarity process after this? That is a subject for another book. Whatever else that other book may have to offer, we believe that in concluding this one we can do no better than to quote Adler's way of summing the matter up:

> From the very beginning the consultant must try to make it clear that the responsibility for his cure is the patient's business. . . . One should always look at the treatment and the cure not as the success of the consultant but as the success of the patient. The adviser can only point out the mistakes; *it is the patient who must make the truth living* [italics added]. (p. 336)

Exercises for Chapter 14.
The Pattern of Basic Convictions and
The Mistaken Ideas
(Use in dyads, small groups, or class, or with clients.)

   In a current or past case of yours, or a case you have studied
or are studying:

1. Using the *sequence* of the client's early recollections (ERs),
   arrange them in a list of key words or phrases.

2. Review the sequence; notice *similarities* among them.

3. Are any ERs juxtaposed to other ERs, revealing *symmetries*?

4. Referring to Figure 8 (p. 403) as providing a way to consider
   antithetical ideas, identify the antithetical convictions unique to
   the client.

REFERENCES

Adler, A. (1963). *The problem child.* New York: Capricorn. (Original work published 1930)

Adler, A. (1964). *The Individual Psychology of Alfred Adler: A systematic presentation in selections from his writings.* (H. L. Ansbacher & R. R. Ansbacher, Eds.). New York: Harper.

Adler, A. (1970). *The education of children.* (E. Jensen & F. Jensen, Trans.). Chicago: Henry Regnery. (Original work published 1930)

Adler, A. (1979). *Superiority and social interest: A collection of later writings* (3rd Rev. ed.). (H. L. Ansbacher & R. R. Ansbacher, Eds.). New York: Norton.

Adler, A. (1980). *What life should mean to you.* (A. Porter, Ed.). New York: G.P. Putnam/Perigee. (Original work published 1931)

Adler, A. (1982). *Cooperation between the sexes: Writings on women and men, love and marriage, and sexuality.* (H. L. Ansbacher & R. R. Ansbacher, Eds.). New York: Norton.

Ahlstrom, S. (1972). *A religious history of the American people.* New Haven: Yale.

Allen, T. (Producer). (1969). *Rudolf Dreikurs: Individual Psychology counseling and education* (Part I). Alexandria, VA: American Association for Counseling and Development. (Film).

American heritage dictionary of the English language (3rd ed.). (1992). New York: Houghton Mifflin.

Bacon, F. (1620). *Novum organum: The new logic or true directions for the interpretation of nature.* [Electronic version]. Retrieved April 6, 2007. http://www.philosophy.Leeds.ac.uk/GMR/hmp/texts/modern/Bacon/Novorg.html

Bartlett, J. (1968). *Familiar quotations* (14th Rev. ed.). (E. Beck & M. Rackliffe, Eds.). Boston: Little, Brown.

Baruth, L., & Eckstein, D. (1981). *Life style: Theory, practice, and research.* Dubuque, IA: Kendall-Hunt.

Belove, P. L. (1980). First encounters of a close kind. *Journal of Individual Psychology, 36*(2), 191-208.

Breit, W., & Hirsch, B. T. (Eds.). (2004). *Lives of the laureates* (4th ed.). Cambridge, MA: MIT.

Burtt, S. (2009, February 15). Saying yes to Ryan. *The New York Times Magazine, 66.*

Dinkmeyer, D., Jr., & Sperry, L. (2000). *Counseling and psychotherapy: An integrated approach* (3rd ed.). Columbus, OH: Merrill.

Dreikurs, R. (1973). *Psychodynamics, psychotherapy, and counseling.* (Rev. ed.). Chicago: Alfred Adler Institute. (Original work published 1967)

Dreikurs, R. (with Vicki Stoltz). (1964). *Children: The challenge.* New York: Hawthorn.

Dreikurs, R., Mosak, H. H., & Shulman, B. H. (1952). Patient-therapist relationship in multiple psychotherapy, I and II. *Psychiatric Quarterly, 26,* 219-227; 590-596.

Eckstein, D., Baruth, L., & Mahrer, D. (1982). *Life style* (2nd ed.). Dubuque, IA: Kendall-Hunt.

Eliade, M. (1959). *Cosmos and history.* New York: Harper.

Ellenberger, H. (1970). *The discovery of the unconscious.* New York: Basic Books/Harper Colophon.

Fausto-Sterling, A. (1993, March/April). The five sexes: Why male and female are not enough. *The Sciences. New York Academy of Sciences, 33*(2), 20-35.

Griffith, J. (1984). Organ jargon. *Individual Psychology: The Journal of Adlerian Theory, Research and Practice, 40*(4), 437- 444.

Griffith, J., & Powers, R. L. (2007). *The Lexicon of Adlerian Psychology: 106 terms associated with the Individual Psychology of Alfred Adler* (2nd Rev. ed.). Port Townsend, WA: Adlerian Psychology Associates. E-book, 2010. www.adlerianpsychologyassociates.com

Klass, P. (2009, September 7). *Birthorder: Fun to debate, but how important?* [Electronic version]. Retrieved October 8, 2010. http://www.nytimes.com/2009/09/08/health/08Klas.html.?em

Mosak, H. H. (1958). Early recollections as a projective technique. *Journal of Projective Techniques, 22,* 302-311.

Mosak, H. H., & Shulman, B. H. (1971). *The life style inventory.* Chicago: Alfred Adler Institute.

Nunez, S. (2011). *Sempre Susan: A memoir of Susan Sontag.* New York: Atlas.

Powers, R. L. (1973). Myth and memory. In H. H. Mosak (Ed.), *Alfred Adler: His influence on psychology today* (pp. 271-290). Park Ridge, NJ: Noyes.

Powers, R. L., & Griffith, J. (1982). Psycho-Clarity. *Individual Psychology Reporter (1),* 1.

Powers, R. L., & Griffith, J. (1986). The big numbers. *Individual Psychology Reporter (4),* 3.

Powers, R. L., & Griffith, J. (2011). *The Individual Psychology client workbook* (3rd Rev. ed.). Port Townsend, WA: Adlerian Psychology Associates. E-book, 2011. www.adlerianpsychologyassociates.com

Saulny, S. (2011, January 30). Black? White? Asian? More young Americans choose all of the above. *The New York Times,* pp. A1, A18, A21.

Shawn, A. (2010, December 5). Phantom twin. *The New York Times,* pp. ST116, ST118, ST122.

Shulman, B. H. (1973). *Contributions to Individual Psychology.* Chicago: Alfred Adler Institute.

Turnbull, L. (2010, October 10). Projects' aim to support bullied gay teens. *Peninsula Daily News*, p. C4.

Untermeyer, L. (Ed.). (1942). *Modern American poetry/Modern British poetry* (Combined 4th and 5th eds.). New York: Harcourt.

Vitello, P. (2006, August 20). The trouble when Jane becomes Jack. *The New York Times,* pp. ST1, ST6.

Walton, F.X. (1980). *Winning teenagers over.* Columbia, SC: Adlerian Child Care Books.

Warsett, G. (2010, September 12). The anatomy of a breakup. *The New York Times,* p. ST6.

Willis, J., & Todorov, A. (2006). First impressions: Making up your mind after a 100-Ms exposure to a face. *Association for Psychological Science, 17*(7), 592-598.

Winthrop, H. (1959). Scientism in psychology. *Journal of Individual Psychology (15),* 112-120.

INDEX

Abbreviations: *Birthorder, BO; Community Feeling, CF; Early Recollections, ERs; Family Constellation, FC; Gender Guiding Lines, GGL; Individual Psychology, IP; Initial Interview, II; Lifestyle, LS; Role Models, RM*

A (General Index follows Adler, A.)

Adler, A.
a real explanation, 397
adaptation, 5-6
Adler, secondborn, 233
Alles kann auch anders sein (G.), 60
antithetical . . . apperception, 286
arrangement, 20, 24
assessment, 7 (*See* LS Assessment)
besmirch a clean conscience, 404
biases, primitive attempts of the child [to orient himself], 401
birthorder, 59
child development, 247
children, 185
client change, 397
collaboration, 6, 19-20
community feeling, an evaluative attitude, 400
    Gemeinschaftsgefühl (G.), feeling for the community, 400
community participation, 6
compensation, 248
concretized fictional goal, 258
conscious and unconscious, 328
context, 5-6
    emphasis on, 28
contribution, 6
correct image, 144
correcting errors, 137
creative interpretation, 246
creative power of the child, 247
depreciation tendency, 295
dethronement, 223

Adler, A. (continued)
diagnosis, 41
ERs, 11, 277-278
ERs, no chance memories, 277
    embeddedness, social, 7, 265
errors in interpretation, 144
every neurotic is partly right, 29
exogenous factor, 66
facts, and one's opinion of facts 264
family constellation, 351
feelings, index to evaluation, 283
    inferiority, 248
fictional goal of life, 258
field of inquiry, 5
firstborn, 232
foremost task of IP, 328
from defeat to victory, 31
goal-directed movement, 61, 277 278
heredity, 245, 247-248, 251-252
    neither heredity nor environment, 246
    shaping of genetic inheritance, 251-252
individual uses experiences in his own creative way in building up his attitude toward life, 246
individual variant, 9, 12, 183, 208, 219, 220, 257, 265
inferior= below=feminine versus powerful=above=masculine, 187
intervention, 6
interview schedule, 27
IP practice, 16

413

Jane S. Griffith and Robert L. Powers, in addition to the present work, are coauthors of *The Lexicon of Adlerian Psychology: 106 Terms Associated with the Individual Psychology of Alfred Adler* (2nd Rev. ed., 2007; e-book, 2010) and *The Individual Psychology Client Workbook* 3rd Rev. ed., 2012; e-book, 2012). They are column editors for the *Journal of Individual Psychology* and are authors of numerous professional articles.

Powers is a graduate of Capital University, Columbus, Ohio (BA), the University of Chicago (MA), and Yale University (MDiv). He is an Episcopal priest. He earned the Certificate in Psychotherapy from the Alfred Adler Institute of Chicago, now the Adler School of Professional Psychology (ASPP), where he is the Distinguished Service Professor of Adlerian Studies in Culture and Personality, *emeritus*. He is a past-president of the Chicago Psychological Association and of the North American Society of Adlerian Psychology (NASAP). He is a Diplomate in Adlerian Psychology and a licensed clinical psychologist.

Griffith is a graduate of Hollins University, Roanoke, Virginia (BA), the University of Maine (MAT), and the Alfred Adler Institute of Chicago, now the ASPP (MA), where she is a professor *emerita*. She is a past-president of NASAP, a Diplomate in Adlerian Psychology, and a licensed clinical professional counselor.

Together they founded the Americas Institute of Adlerian Studies (1982-1992), the first Adlerian training program approved for CE by APA and NBCC. They conduct the four-course *Certificate Program in Adlerian Psychology* for the Puget Sound Adlerian Society, and lecture in the United States and abroad. In 2011, they jointly received NASAP's *Lifetime Achievement Award*.

Powers and Griffith are married to each other and live in Port Townsend, Washington. CONTACT: www.adlerianpsychologyassociates.com or info@adlerpsy.com

CPSIA information can be obtained at www.ICGtesting.com
Printed in the USA
BVOW04s0814240913

331989BV00001B/13/P